SCIENTISTS' EXPERTISE AS PERFORMANCE: BETWEEN STATE AND SOCIETY, 1860–1960

History and Philosophy of Technoscience

SCIENTISTS' EXPERTISE AS PERFORMANCE: BETWEEN STATE AND SOCIETY, 1860–1960

EDITED BY

Joris Vandendriessche, Evert Peeters and Kaat Wils

LONDON AND NEW YORK

First published 2015 by Pickering & Chatto (Publishers) Limited

2 Park Square, Milton Park, Abingdon, Oxfordshire OX14 4RN
52 Vanderbilt Avenue, New York, NY 10017

Routledge is an imprint of the Taylor & Francis Group, an informa business

First issued in paperback 2020

BRITISH LIBRARY CATALOGUING IN PUBLICATION DATA

Scientists' expertise as performance: between state and
society, 1860–1960. – (History and philosophy of technoscience)
1. Science – Social aspects – History. 2. Expertise – Social aspects – History.
I. Series II. Vandendriessche, Joris, editor. III. Peeters, Evert editor.
IV. Wils, Kaat editor.
303.4'83'09-dc23

ISBN-13: 978-1-8489-3527-3 (hbk)
ISBN-13: 978-0-367-59980-5 (pbk)

Typeset by Pickering & Chatto (Publishers) Limited

CONTENTS

ACKNOWLEDGEMENTS

This volume is the result of the work of a group of historians who came from different subfields to the history of expertise. Its foundations were laid during the international conference *Between Autonomy and Engagement. Performances of Scientific Expertise, 1860–1960* which took place in Leuven, Belgium, from 21 to 23 May 2012. The conference was organized by the Research Group Cultural History since 1750 of the University of Leuven, and the Leuven Interdisciplinary Platform for the Study of the Sciences (LIPSS). It was financed by the Research Foundation Flanders (FWO), the Faculty of Arts of the University of Leuven, the Leuven OJO fund for the support of young researchers and the Academische Stichting Leuven. We would like to thank all of the participants in the conference and in the LIPSS reading group on scientific expertise. Their reflections regarding the concept of 'expertise' have been invaluable in the further development of the volume. We are grateful to the contributors to this volume for their continuing intellectual engagement and their valuable suggestions during the publishing process. We would like to express our gratitude to all those involved in the process of creating, producing and finishing this book.

LIST OF CONTRIBUTORS

Katja Bruisch is a research fellow at the German Historical Institute in Moscow. Her fields of research include the history of science and knowledge with a focus on the history of economic thought. She also has a strong interest in the rural and environmental history of late Imperial and Soviet Russia. Her publications include *Als das Dorf noch Zukunft war. Agrarismus und Expertise zwischen Zarenreich und Sowjetunion* (Köln et al.: Böhlau, 2014); *Bol'shaia voina Rossii: Sotsial'nyi poriadok, publichnaia kommunikatsiia i nasilie na rubezhe tsarskoi i sovetskoi èpokh* (Moscow: Novoe Literaturnoe Obozrenie, 2014), co-edited with N. Katzer, and 'Historicizing Chaianov. Intellectual and Scientific Roots of the Theory of Peasant Economy', in D. Müller and A. Harre (eds), *Jahrbuch für Geschichte des ländlichen Raums* (Special Issue: *Transforming Rural Societies. Agrarian Property and Agrarianism in East Central Europe in the Nineteenth and Twentieth Centuries*, 2010), pp. 96–113.

Raf de Bont is Assistant Professor at the Faculty of Arts and Social Sciences at Maastricht University. His research interest concerns the history of science in the nineteenth and twentieth centuries. He has published particularly on scientific ecology and nature protection, the relation between laboratory and field biology, the interaction between the social and the life sciences, and the representation of science (and scientists) in culture at large. His latest book, *Stations in the Field: A History of Place-Based Animal Research, 1870–1930*, will be published in 2015 with the University of Chicago Press (Chicago, IL).

Margo De Koster is part-time Professor at the Criminology Department of the Vrije Universiteit Brussel, Lecturer in Historical Criminology at the Vrije Universiteit Amsterdam and Vice-Coordinator of the Belgian research network 'Justice & Populations: The Belgian Experience in International Perspective, 1795–2015' (Belgian Science Policy Office, 2012–17). She conducts historical-criminological research on juvenile delinquency and juvenile justice, girls and women in criminal justice, urban policing and transgressive uses of urban public space. She is a member of the editorial board of the journal, *Crime, History & Societies / Crime, Histoire et Sociétés*.

David Freis is a graduate student in the Department of History and Civilization at the European University Institute (EUI) in Florence, Italy. His research interest concerns the history of medicine, psychiatry and the psychological disciplines, central European history and the history of the interwar period. His recent publications include 'Die "Psychopathen" und die "Volksseele": Psychiatrische Diagnosen des Politischen und die Novemberrevolution 1918/1919', in H.-W. Schmuhl and V. Roelcke (eds), *'Heroische Therapien': Die deutsche Psychiatrie im internationalen Vergleich 1918–1945* (Göttingen: Wallstein, 2013), pp. 48–68, and 'Homosexualität und Männlichkeit im Spannungsfeld von Justiz, Psychiatrie, Militär und Adel: Ein Fall aus der forensischen Militärpsychiatrie des Ersten Weltkriegs', *kultur & geschlecht*, 7 (2011), pp. 1–29.

Graeme Gooday is Professor of History of Science & Technology and Head of the School of Philosophy, Religion and History of Science at the University of Leeds. His main interests are in the social history of technology, particularly the electrical technosciences. At present his research concerns the historical cultures of patenting; the politics of telecommunications usage (especially in the First World War), and the controversies over early hearing assistive technologies. His books include: *Patently Contestable: Electrical Technologies and Inventor Identities on Trial in Britain* (Cambridge, MA: MIT Press, 2013), with Stathis Arapostathis; *Domesticating Electricity: Technology, Uncertainty and Gender, 1880–1914* (London: Pickering & Chatto, 2008); *Physics in Oxford 1839–1939: Laboratories, Learning and College Life* (Oxford: Oxford University Press, 2005), co-edited with Robert Fox; and *The Morals of Measurement: Accuracy, Irony and Trust in Late Victorian Electrical Practice* (Cambridge and New York, NY: Cambridge University Press, 2004).

Frank Huisman is Professor in the History of Medicine, affiliated to the Julius Centre of the University Medical Centre Utrecht and to the Descartes Centre for the History and Philosophy of the Sciences and the Humanities of Utrecht University. He has published on medical historiography, quackery and the cultural authority of medicine. His publications include *Stadsbelang en standsbesef. Gezondheidszorg en medisch beroep in Groningen 1500–1730* (Rotterdam: Erasmus Publishing, 1992), a local case study of early modern Dutch health care; *Locating Medical History. The Stories and their Meanings* (Baltimore, MD: The Johns Hopkins University Press, 2004), co-edited with John Harley Warner; and *Health and Citizenship. Political Cultures of Health in Modern Europe* (London: Pickering & Chatto, 2014), co-edited with Harry Oosterhuis. He is now working on a book exploring the transformation of the Dutch health care system between 1880 and 1940. Huisman is one of the initiators of 'Science in Transition', a movement which believes that science and the university are in need of fundamental reform (http://www.scienceintransition.nl/english).

Jennifer Karns Alexander holds a joint appointment at the University of Minnesota in both the Program in History of Science, Technology, and Medicine and the Department of Mechanical Engineering. She is a historian of technology specializing in modern technological culture. Her current research area is technology and religion in the modern world. Her publications include *The Mantra of Efficiency: From Waterwheel to Social Control* (Baltimore, MD: The Johns Hopkins University Press, 2008).

Martin Kohlrausch is Associate Professor of European History at the University of Leuven. His fields of research comprise the history of architects, political history and history of mass media. His publications include *Building Europe on Expertise. Innovators, Organizers, Networkers* (London: Palgrave Macmillan, 2014), with Helmuth Trischler; 'Postcatastrophic Cities', special issue of the *Journal of Modern European History*, 9:3 (2011), co-edited with Stefan-Ludwig Hoffmann; and *Expert Cultures in Central Eastern Europe. The Internationalization of Knowledge and the Transformation of Nation States Since World War I* (Osnabrück: Fibre, 2010), co-edited with Katrin Steffen and Stefan Wiederkehr.

Per Lundin is Associate Professor of Economic History at Uppsala University Centre for Science and Technology Studies. He received his PhD in History of Technology from the Royal Institute of Technology in Stockholm. His published dissertation, *Bilsamhället: Ideologi, expertis och regelskapande i efterkrigstidens Sverige* (Stockholm: Stockholmia Förlag, 2008) (The Car Society), was awarded the Johan Nordström and Sten Lindroth Prize for outstanding scholarly work in the history of ideas. His publications further include *Computers in Swedish Society: Documenting Early Use and Trends* (Dordrecht: Springer, 2012) and, co-edited with Niklas Stenlås and Johan Gribbe, *Science for Welfare and Warfare: Technology and State Initiative in Cold War Sweden* (Sagamore Beach, MA: Science History Publications, 2010).

David Niget is Assistant Professor in the Department of History at the University of Angers. He has a PhD from the Université du Québec à Montréal and was a postdoctoral fellow at the University of Louvain. His primary research topic is juvenile delinquency and youth culture, expertise and child guidance, moral panics and risk. He is currently preparing an international research project on the history of children's rights and citizenship. He is a member of the editorial board of the *Revue d'histoire de l'enfance irrégulière / Journal of the History of 'Irregular' Childhood*.

Evert Peeters is a cultural historian at the Research Group Cultural History since 1750 at the University of Leuven. In 2008, his doctoral dissertation on *Lebensreform* and cultures of the body in Belgium, *De beloften van het lichaam. Een geschiedenis van de natuurlijke levenswijze, 1890–1940* (Amsterdam: Bert Bakker, 2008) was published. With Bruno Benvindo, he published a monograph on the changing memories of the Second World War in Belgium: *Scherven van de oorlog: de strijd om de herinnering aan de Tweede Wereldoorlog, 1945–2010* (Antwerp: De Bezige Bij, 2011). He co-edited the volume *Beyond Pleasure: Cultures of Modern Asceticism* (New York, NY: Berghahn, 2011) with Kaat Wils and Leen Van Molle. His current research focuses on psychological expertise and the performance of labour in the early twentieth century.

Niklas Stenlås is Associate Professor of Economic History at Uppsala University Centre for Science and Technology Studies. His publications include *Den inre kretsen: Den svenska ekonomiska elitens inflytande över partipolitik och opinionsbildning 1940–1949* (Lund: Arkiv förlag, 1998), on the influence of the Swedish economic elite over political parties and opinion moulding, and a number of articles on the history of Sweden's political parties, Sweden's military-industrial complex and the deprofessionalization of civil servants. Together with Per Lundin and Johan Gribbe, he has edited the volume *Science for Welfare and Warfare: Technology and State Initiative in Cold War Sweden* (Sagamore Beach, MA: Science History Publications, 2010).

Martin Theaker is a doctoral research candidate at Trinity Hall, University of Cambridge. His thesis concerns the role of civil atomic energy in Britain's relations with Europe during the 1950s, and the growth of expert influence within London's policymaking processes. His other research interests include the postwar interaction between Britain, the United States and the Commonwealth in the atomic energy field, as well as the role played by nuclear power in redefining Britain's energy economy after 1956.

Joris Vandendriessche is a postdoctoral researcher at the Research Group Cultural History since 1750 at the University of Leuven. His doctoral research focused on scientific culture in nineteenth-century medical societies in Belgium. His research interests comprise the history of the medical sciences, public health and health care in the nineteenth and twentieth centuries. Recent publications include 'Setting Scientific Standards. Publishing in Medical Societies in Nineteenth-Century Belgium', *Bulletin of the History of Medicine*, 88:4 (2014), pp. 626–653, and together with Kaat Wils, 'Een traject van onderhandeling. Hygiënisme als wetenschap, Antwerpen, 1880–1900', *BMGN – The Low Countries Historical Review*, 128:3 (2013), pp. 3–28.

Kaat Wils is Professor in Contemporary Cultural History and head of the Research Group Cultural History since 1750 at the University of Leuven. Her dissertation on the history of positivism and intellectual culture in Belgium and the Netherlands was published in 2005: *De omweg van de wetenschap: het positivisme en de Belgische en Nederlandse intellectuele cultuur, 1845–1914* (Amsterdam: Amsterdam University Press, 2005). Her research focuses on the history of the humanities and the biomedical sciences, and the history of gender, corporality and sexuality. She has been editor of several volumes on the history of the body and medical cultures, including: E. Peeters, L. Van Molle and K. Wils (eds), *Beyond Pleasure: Cultures of Modern Asceticism* (New York, NY: Berghahn, 2011).

LIST OF FIGURES

INTRODUCTION: PERFORMING EXPERTISE

Joris Vandendriessche, Evert Peeters and Kaat Wils

For the Russian agronomist Aleksei Doiarenko, scientific ambition, social engagement and state administration had always been closely entangled.[1] As a young student at the Agricultural Institute in Moscow around 1900, he started lecturing the farming population about agricultural modernization – a topic he embraced in the hopes of politically emancipating the Russian countryside. After obtaining a professorship at the Agricultural Academy, he quickly became a leading voice in the emerging field of agricultural development. In the same period, he became absorbed by state administration. During the early 1920s, he entered the People's Commissariat of Agriculture where he was involved in the first attempts at large-scale agricultural planning in Soviet Russia. For Doiarenko, the rise to expert fame implied a complex interweaving of social roles. As the scholar in agronomy gradually evolved into a popular educator and an administrator, he needed to navigate between very different audiences, all of which seemed to formulate their own (academic, popular and political) prerequisites. These prerequisites often changed – and in revolutionary Russia they could change overnight. As the Great Purges turned upside down the preconditions of expert recognition, and 'pre-revolutionary', imperial experts were easily replaced by a stratum of newly educated and thoroughly communist technicians, Doiarenko was stripped of expert authority, only to be rehabilitated posthumously during the Khrushchev era. Negotiating cautiously between state, science and society, Doiarenko finally faced the loss of political patronage – a loss that could not be compensated by his previously accumulated academic and popular respectability.

The fate of Doiarenko is exemplary of the aura and the fragility of (academic) expertise in modern and late modern policymaking. For all the obvious successes of expert politics since the second half of the nineteenth century, the 'expert society' that may have materialized because of these successes was never fully within the hands of the presumed experts themselves. For at least a decade now, scholars in the history of science and in science studies (often animated by philosophers and sociologists of science), have argued exactly that. From agriculture to public health policies, from experimental to statistical method, in liberal-capitalist as well as in 'totalitarian' politics, expertise continuously expanded into new fields of the social fabric, yet always remained a socially constructed, inherently unstable form of authority, as it sought for simultaneous recognition within and outside the academy, in between state and society.[2] And as the 'expert society' has gradually replaced the notion of 'technocracy' in scholarly discourse, this instability of expertise has been laid bare through a healthy reassessment of individual expert careers, of individual and group-like modes of negotiation and survival, of expert respectability and political success.[3] In this volume, we wish to build upon these reassessments, through reformulating the 'expert experience' as a set of specific expert 'encounters' with the state and society, encounters which resulted in a renegotiation of the boundaries between these entities. It may be true that structural changes in western societies such as the expansion of state power, the ever closer interweaving of state and society and the growing complexity of policymaking altogether, have cleared the pathway for the advent of expert authority. Yet expertise only materialized through the *performances* of experts, who navigated continuously and carefully between the changing boundaries of state and society. These performances form the subject of this book.

In focusing upon expert performances between state and society, this volume builds upon two specific scholarly trends that have gained momentum during the last decade. Among historians of science, in the first place, the growing interest in 'expert societies' – both as socially engaged networks of scientists and as western societies in which these experts' influence seemed to expand – clearly tapped into the broadly felt ambition to reformulate their own object as the study of all sorts of encounters between science and society, and to move beyond the study of discipline formation per se.[4] Whereas these ambitions may have materialized most clearly in the growing interest in the 'cultural' aspects of discipline formation itself, they have also incited historians to study the 'scientization' of ever wider problems of social life from the nineteenth century onwards. Departing from the long-held belief that a scientific gaze could exemplify a social regime in itself – an approach which has reigned most forcefully among historians of medicine – historians of science have changed focus to the often localized interactions between scientists, public opinion and political establishment on the one hand, and the interplay between the power of scientific discourse and the attainment of social respectability by scientists on the other. Especially with regards to the social sciences, Lutz Raphael's famous thesis about the 'scientization of the social' has summed up most clearly that agenda. For Raphael, the rise of expert influence developed simultaneously with the pervasion of social methodologies within which the legal capacities, the ability to 'produce efficiently' and even the ability of happiness of social groups and individuals have been assembled since the nineteenth century. At the same time, however, Raphael has argued that these methods, whether they were medical or statistical, emerging from military and labour psychologies or embedded within the sociology of public opinion surveys, always met with legal borders and with social resistance. Therefore the 'scientization of the social' should not be equated with the deployment of 'disciplinary' power, but with the continuous negotiation between political, economic, social and cultural contexts.[5]

Yet whereas historians of science have increasingly questioned science-society interactions during the last two decades, 'expertise' has gained an even stronger totemic status, in the second place, within science studies. In this interdisciplinary field as well, scholars have redefined the 'technocratic' negotiation between state power, the public sphere and academic authority in terms of expert performances.[6] Rather than considering expertise as a passive outcome of this negotiation, many scholars have stressed to what extent experts brought about this negotiation and therefore shaped these encounters, alongside forms and beliefs that originated in between science, state and society. As the aim of many scholars in this field has been to design models of interaction between scientists and policymakers, the strategies used by experts to convince such an audience were crucial. In their efforts, the work of the American sociologist Erving Goffman, who used the metaphor of the theatre to study how individuals presented themselves in everyday interactions, proved influential.[7] Applying Goffman's theories in his *Science on Stage,* Stephen Hilgartner studied the apparatus through which science advice gained credibility, looking, for example, at the self-representation of science advisers and the reception of their advice. Hilgartner showed how science advisers tried to inspire confidence among their audiences by displaying trustworthiness, competence and integrity through their rhetoric and comportment.[8] Wiebe Bijker, similarly inspired by Goffman's metaphor, made a division between expertise's 'front stage' – the presentation of expert knowledge as a finished product – and 'back stage' – the process of producing expert knowledge. Such a division can equally be viewed as an effort to identify the different audiences to which experts are necessarily addressing themselves, and the inherent 'cleavage' that punctuates their performance.[9]

In this book, we aim to mobilize these gains from both the history of science and the science studies in order to further clarify the effectiveness of experts – and their expertise – within modern knowledge societies. For if historians and sociologists of science have rightly refigured the frameworks within which expertise has come to dominate contemporary societies, this effectiveness itself remains to be fully grasped. In other words: if expert authority was the outcome of complex encounters between scientists, society and the state rather than the reflection of a distinct technocratic structure, it remains to be investigated why experts emerged so remarkably powerful from some of these encounters, whereas they were so easily defeated in others. And if expert performances – in the theatrical sense – may have contributed so much to the establishment of expert authority, then what sorts of performances underlie the relative success of Doiarenko and others, and the failures of so many of his competitors? What seems to be at stake here is a reformulation of expert performances under the denominator of their effectiveness or their ability to shape modern societies alongside the assumptions that were structuring their particular 'scientization' of reality. In science studies, this effectiveness has often been defined as 'performativity', or the resorting of external effects.[10] In other words: the effectiveness of experts' role-playing before a particular audience and their ability to bring about change in the outside world are intrinsically linked up.

In his *Science on Stage* mentioned above, Hilgartner has mainly focused on the first aspect, when analyzing the extent to which experts succeeded in convincing their ever expanding audiences of their own authority, and managed to secure positions of power on the basis of those performances. With Hilgartner, we aim to see the effectiveness (or 'performativity') of expertise above all as a problem of embodiment – an embodiment by experts of a specialist role, and a set of scientific and social ideals connected with it. Yet, as many other sociologists of science have argued, this role-playing prefigured the transforming power of particular sorts of specialist knowledge in the outside world. As the British sociologist of science Donald MacKenzie has recently argued with regards to the role of economic expertise in the making of financial markets, modern experts not only endorse themselves with guidelines and instruments in order to convince the outside worlds of their powers in a theatrical manner, they do effectively transfigure social realities with the help of these instruments. Whereas, in his wordings, the performativity of expertise sometimes comes in very weak forms, in which case specialist knowledge is merely used to navigate complex social realities, effective expertise gradually becomes an engine of change in modern societies. In the strongest forms, like in Doiarenko's, social realities are increasingly shaped alongside the patterns of specialist knowledge.[11] In this book, we aim to investigate anew how, in the history of expertise, these 'weak' and 'strong' forms of expert effectiveness have been correlating with each other. Or in other words: how the encounters and role-playing of modern experts have made the expert's society first thinkable, and then do-able.

Agency and Audience: The Innovations of Early Modern Expertise Scholarship

When studying expert performances and expert effectiveness, it may seem audacious to have historical scholarship tied in with the conceptual frameworks of the science studies. And even though historians have increasingly shifted towards broad histories of scientization and expertise, the performances within which much of this expertise seems to have materialized, have hitherto been taken up by them in rather diffuse manners. Also, historians have been less interested in modelling expert interaction than scholars in the science studies.[12] And yet, with regards to expert embodiments on the one hand and the effectiveness of expertise on the other, many intuitions from the science studies have tacitly found their way to historical scholarship. These intuitions have not pressed historians to 'model' expert interactions, but rather helped them deepen their insight into the practices and procedures that experts have developed when interacting with the state and society. In particular with regards to early modern societies, recent scholarship has intensely discussed the expert performances of astronomers and engineers, optics and fortress builders as they evolved centuries before the 'scientization of the social', or the contemporary expert performances that are being discussed in the science studies.

The sixteenth to eighteenth centuries saw a proliferation of advisory practices as performed by different sorts of savants, whereas they also saw the first emergence of the concept 'expert (-witness)', within the context of the courtroom.[13] Small-scaled as these expert practices may have been, fully embedded within an interpersonal exchange between adviser and patron, and untainted by modern divisions between front-stage and back-stage, they nonetheless present the raw birth of a particular engagement of the expert with his audience. That is why Eric Ash, for instance, presents the early modern period as the first 'laboratory of expertise'. What came out of this laboratory were performances, rather than material effects; functions, let alone professions. As Ash convincingly points out, the early modern expert lacked most of the formal institutions that could provide experts with 'external' credentials, such as modern universities, government bureaus or professional organizations. Therefore, Ash continues, early modern expertise typically emerged as a cautiously crafted relationship of the expert with the audience he wished to convince of his credibility. Apart from the material effects of the expert's intervention (the astronomic calculation, the fortress or the bridge), the expert status seemed to depend solely on the public's willingness to recognize the expert as the main author of these effects, and thus on the expert's ability to present his competences in a compelling manner.[14]

Going further, historians of early modern expertise have demonstrated to what extent the agency of experts originated in their capacity to mediate between plural audiences. The concept of the 'expert mediator', developed by Ash in his study of expertise in Elizabethan England, places this interplay between experts and their (successive) audiences centre-stage. For Ash, the experts of the Elizabethan state constantly mediated between their patrons, the central administrators and the objects these patrons wanted to control.[15] Such mediation required particular skills and strategies. Audiences, comprised of state officials, needed to be convinced of the validity of their expert knowledge. Such a necessity of demonstrating their skills and highlighting their capabilities was crucial for the early modern expert, whose legitimacy depended entirely on the support of these state officials.

The link between this dependence and expert performances is, for example, shown by Andre Wakefield in his study of the discussions between the Duke of Hanover and Gottfried Wilhelm Leibniz on the exploitation of the Harz silver mines. Wakefield shows how Leibniz trusted his connections to scientific academies and the rhetorical superiority of philosophical knowledge over the practical experience claimed by mining officials, to present himself as the leading expert on mining.[16] Again, the expert emerges as a mediator between the available knowledge and those wanting to use this knowledge for political or economical purposes. Ursula Klein has recently described these negotiations from the viewpoint of material culture as mediations 'between consumers and their goods'.[17] In early modern studies, the expert mediator has indeed formed a fruitful concept to unravel the way experts presented themselves as necessary links in the expansion of government power into new social areas by playing into desires and interests of their patrons.

Not surprisingly, the engagement of early modern experts with different audiences at a time, each with their own prerequisites, also helps to explain the plural character of expert roles themselves. As several historians have demonstrated, early modern experts did not form a distinct group or class in society. The variety of experts rather reveals that expertise should be viewed as the taking up of expert roles, which could be done effectively by all sorts of men. Apart from the early modern savants and scholars active in academies who acted as experts when asked to provide advice to the government,[18] early modern historians have also pointed to a wide range of experts outside of scholarly circles. What made these men into 'experts' was their combining of learned knowledge with hands-on experience, 'borrowing skill, language, and explanations from both the artisanal and the scholarly worlds'.[19] Engineers, anatomists and physiologists, as well as mining officials or assayers, and many others could fit this category. What united them was their ability to blend different forms of knowledge to meet new niches of an expanding market of knowledge. The eighteenth-century 'artisanal-scientific expert' forms a typical example of these hybrid figures. Supported by the mercantilist state and keeping close ties with both industry and the world of academies, these men – chemists, botanists and engineers – took up expert roles that developed together with new state bureaucracies.[20] These diverse examples show that being a successful expert meant identifying the right audiences and presenting one's knowledge in such a way that it met the needs of these audiences. In these studies in early modern history, expertise appeared as something 'intermediary', which could only be investigated by looking at the audiences and spaces of expertise.

Expertise and Scientization Since Modernity

For historians of modernity and beyond, these innovations of early modern expertise prove to be of great importance. In fact, experts' role-playing and continuous engagement with different audiences continued well beyond the early modern period. As has been implicitly argued in many historical studies since Roy Macleod's collection *Government and Expertise* in 1988, modern experts continued to invent their expertise themselves by responding to the needs of their audiences – the newly emerging administrative elites of liberal-capitalist societies – and thus to a large extent shaped their own fields of activity.[21] Yet the expansion of government intervention in ever more areas of social life dramatically changed the conditions within which these traditional advisory roles were being performed. Roughly between 1860 and 1960, opportunities for state-backed expert performance rose sharply as the different fields of government intervention were to be shaped and conceptualized, whereas the public and private institutions of technoscience transformed traditional expert crafts and academic disciplines.[22] At the same time, expert performances became loaded with the scientific ideals, the particular visions of the state and the ideological frameworks of the society in which service they believed to operate. As the scope of these expert performances expanded from the second half of the nineteenth century onwards, its effectiveness seemed to grow as well. This rise of modern 'technoscientific' expertise was a two-sided process. State expansion clearly generated an increasing demand for technical and scientific knowledge. But at the same time, processes of professionalization and specialization expanded the 'supply side' of expertise and legitimated expert interventions in new areas. The result was a transformation of the early modern advisory expert role into a more complex form of expertise, which was also legitimized in new ways as science became ever more institutionalized in universities and academic disciplines.[23]

These structural changes may be beyond discussion – the question remains, however, how these shifts are best defined. As Thomas Broman recently emphasized, the place of the expert within them remains to be fully grasped – both as outcome and as inventor of newly emerging interactions between science, society and the state.[24] By reformulating fundamental changes in western societies in terms of an increasing effectiveness of expertise, this volume draws attention to experts as individual actors who constantly reshaped the boundaries between science, state and society. In particular, three fundamental features of modern technoscientific expertise are taken into account. The first feature of this kind concerns the scientific grounding of expertise. Certainly, early modern experts often claimed to possess a theoretical, abstract knowledge that was more objective than the technical skills and experience of craftsmen, which were said to be based in personal interests.[25] Modern experts, however, increasingly referred to the scientific basis of their knowledge to formulate similar claims of objectivity. By stressing the use of scientific methods, and applying scientific terminology, they distinguished themselves from other players in the field.[26] Such claims were strengthened by new institutional affiliations. Modern experts were not only trained at the universities, their expert knowledge was also legitimized by professional organizations, disciplinary communities such as specialized scientific societies and new government commissions.[27] Support of public officials, bureaucrats and political leaders was therefore no longer the only way for experts to gain formal credentials. As a result, the early modern savants, who had provided expert advice at the request of the government, were gradually replaced by an army of new scientific specialists.

Simultaneously, increasing effectiveness of expertise also resulted in growing political and institutional embeddedness of experts – a second feature that we aim to discuss. Their intense affiliations with government institutions made modern experts different from their eighteenth-century predecessors, who functioned within the context of government-funded academies. Modern experts became an integral part of power structures, assuming full-time positions within state infrastructure. Within these power structures, they took up leading roles as decision-makers, organizers and managers. The modern expert could become a statesman and not only legitimize, but also have a profound impact on policymaking. Yet, this shift also produced new problems of independence and credibility. How could experts provide neutral advice to government officials, so critics argued, when they were on the payroll of these same men? Again expertise required a careful balancing between the opportunities of working in government service, and the fragilities that emerged together with the intensified collaboration with the state.[28]

Finally, the increasing effectiveness of expertise is to be situated in a growing entanglement of expertise and political ideologies. In short, together with experts' beliefs in the many blessings of modern science, their expertise also became more grounded in ideology. The very conviction that science could carry through social changes, that scientific development could function as the driving force of social improvements, distinguished modern experts from their eighteenth-century predecessors.[29] Such an ideology was already present among nineteenth-century experts, but these views radicalized in the twentieth century as the opportunities for expert intervention augmented. Public health experts and psychiatrists, for example, but also experts in labour division and educational experts increasingly engaged with politics in an attempt to carry their plans for social change into effect. Experts thus evolved from advisers to reformers. Even though they claimed to transcend party politics by implementing scientific views on social problems, experts increasingly became part of politically inspired reform movements. This paradox illustrates how modern expertise was closely connected to the changing position of science in society. The 'scientization' of politics, based on a shared positivist framework, indeed also meant that experts themselves became players in the political arena.[30]

Expert Encounters

The structure of this book reflects the performative perspective to the history of modern scientific expertise. The four parts represent a general pattern of expert performances: searching for audiences, convincing them, engaging with the state and (re)shaping the social and political objects under expert scrutiny. In that manner, we aim to develop a sample card of expert encounters, within which different degrees of expert effectiveness will be contextualized anew. Of course, these encounters complexified over time: specialist roles diversified as audiences broadened in whose service these roles were performed, and the sets of scientific and social ideals connected with these performances multiplied likewise. As the chapters in this book will demonstrate, expert effectiveness was mediated simultaneously by the leverage of expert audiences, by the 'hospitality' of the specific field of social interaction that experts sought to exploit and by the perceived usefulness, objectivity and reliability of the technical crafts and academic disciplines from which experts derived their specialist knowledge. In other words: the (socio-political) empowerment of expert performances by specific audiences also tied in with the effectiveness of a particular expert gaze. In a Foucauldian sense, the power that evolved from specific interactions between science, state and society always intertwined with the disciplinary creed of specialist knowledge that put ever more fields of social reality under scrutiny. And whilst expert effectiveness originated from the vividness of the conversation between experts and their audiences, this same 'effectiveness' often came down to the objectification of the social realities they studied and shaped. In this manner, expertise itself became the watershed between the limited audiences in whose service experts operated, and the 'objectified' social realities they exploited at the service of these very audiences.

These general patterns of expert performances will be developed from a mostly 'European' perspective. While some chapters analyse expertise in particular national contexts, others take a broader international perspective or focus on specific international spaces of expert performance such as scientific conferences. All of them, however, engage with expertise as a European phenomenon. As Martin Kohlrausch and Helmuth Trischler have recently argued, the development of technoscientific expertise since the second half of the nineteenth century went hand in hand not only with the expanding infrastructure of the new nation states, but also with an increase in the exchange of knowledge on the international level by means of exhibitions and conferences.[31] 'Experts' nationalism and their international mindsets', Kohlrausch and Trischler argue, 'can be fully understood only in reference to one another'.[32] If such features can be considered typically 'European', their occurrence was not limited to the traditional European geography. As the chapters in this book show, similar expert performances took place in the United States and Russia.

In the first part of this volume, *Setting the Scene. Experts and their Public*, attention is paid to the variety of settings in which expertise was performed. The fields of ecological, electrical and medical expertise in the late nineteenth and early twentieth century cannot be understood, so the authors argue, without looking at the international conferences where experts addressed an audience of scientists, the meetings of learned societies where they spoke to their colleagues, or the lecturing halls in which they addressed a popular audience. Although the authors pay attention to the singularities of each of these settings and disciplines, similarities in the construction of expert authority do come to the fore. The scientific claims made by zoologists, electrical experts or public health professionals all functioned as markers that allowed distinguishing them from other aspiring experts, such as administrators, technicians or writers who popularized science. The importance of institutional affiliation, not so much to governmental commissions, but to established institutions that could support the scientific claims on which their authority rested, also stands out. By disentangling such mechanisms of expert authority, each of these contributions presents expertise as a field in the making, which developed 'bottom-up' rather than 'on demand'.

Graeme Gooday examines the performances of three ethnically diverse electrical experts in the settings of the late nineteenth-century experts' witness box in the court room and the lecturing hall. Their careers show, as Gooday concludes, that to become an expert in electricity, not so much mainstream ethnicity was required, but rather performative capacity with the spoken word and institutional affiliation through patrons. An institutional basis was also of importance to aspiring experts in the field of public health. Membership of scientific medical societies, as Joris Vandendriessche shows, formed a way of strengthening one's expert authority. By analyzing the expert performances in two of these societies in nineteenth-century Belgium, Vandendriessche shows how the scientization of the field of public health occurred through reviews and debates in which scientific studies were distinguished from popular, administrative and philanthropic writings. Such delineating practices were also essential to the efforts of ecological experts in the early twentieth century. By examining the international conferences on nature protection, Raf de Bont shows how the authority of zoologists was co-constructed with the image of nature's internationality. This latter image proved an important rhetorical tool in the scientization of the field of nature protection, which led to the rise of zoologists as experts at the expense of ornithologists, colonial administrators, foresters and hunters.

The second part of this volume, *Science as a Belief. Experts and Social Reform*, consists of contributions that scrutinize the messages used by experts to convince their audiences. These messages were not only ways of displaying scientific skills; they also stirred enthusiasm, embodied political ideals and could even touch upon utopian beliefs. Such ideological narratives, as the contributions in this part show, were never homogeneous. In the fields of engineering, psychiatry and agriculture in the early twentieth century, a varied range of political views could be found among experts. This ideological multiformity certainly testifies to the particularity of each of these fields, but it also hints at a more general finding regarding the relation between expertise and politics. As Martin Kohlrausch argues in his contribution on experts in interwar Europe, expertise and politics became strongly intertwined in the early twentieth century. To study this intertwinement, Kohlrausch calls for more attention to the personal trajectories of experts, which can lead to a reassessment of the major turning points in the history of expertise.

The two other contributions in this part respond to this call. David Freis analyses the expert roles taken up by German and Austrian psychiatrists, which were shaped to a large extent by their experience of the First World War. Freis traces, more specifically, their attempts to join conservative political circles and infuse political discourse with medical terminology. By presenting themselves as the 'doctors of the nation' and by 'diagnosing the revolution', Freis concludes, these psychiatrists were able to establish themselves as socially influential professionals. Katja Bruisch equally examines the entanglement of individual trajectories with macro-political developments. In her contribution on agricultural experts in early twentieth-century Russia, Bruisch shows the manoeuvring of experts within rapidly changing political settings. The career of the agricultural scientist Doiarenko, in particular, shows the flexibility of scientific expertise when it comes to inspiring social reform, but also its fragility, as its social relevance was almost entirely based on the patronage of certain decision-makers.

The contributors to the third part of this volume, on *Diplomatic Strategists. National Government and Expert Ambitions*, discuss a specific type of expert encounter. They examine the interaction of experts with an audience of state officials that emerged together with the growing institutional embeddedness of expertise in the post-Second World War welfare state. Per Lundin and Niklas Stenlås demonstrate how a group of Swedish experts – architects, economists, engineers, planners and scientists – were successful in presenting themselves as 'apolitical professionals' and worked their way to leading positions in government service. Lundin and Stenlås consider these experts as 'reform technocrats' and stress their undervalued role in the development of the post-war Swedish welfare state. Martin Theaker focuses on the interaction between state officials and atomic experts in post-war Britain. Theaker extends the definition of the 'scientist-diplomat', which has been used so far in the context of international scientific exchange, by including the domestic mediation of experts between the worlds of politics and science. In both contributions the active role of experts as agents in the expansion of government services is stressed.

In the final part, *Objectification. Expertise and its Discontents*, the contributors reflect upon the relationship between authority and power in expert performances. When discussing this relation, the science studies have typically privileged problems of trust over matters of discipline. Especially with regards to the perceived crisis of legitimacy in contemporary expert societies, sociologists of science have depicted expert authority as the outcome of a conversation that eventually goes awry if increasingly demanding audiences finally outstretch the adaptability of particular groups of experts.[33] For many in the science studies, the sorts of interactions that expert performances entail are defined as the sequences of an ongoing (and mutual) exchange with ever greater and more 'democratic' audiences. In his contribution, Frank Huisman analyses the history of medical expertise in the Netherlands from a similar conversational perspective. Huisman explains the ambiguous fate of medical expertise from its growing inability to mediate between these new, different and more demanding audiences on the one hand, and the scientific sphere on the other. In his comparison of three debates on medical legislation in the Netherlands – in the 1860s, the 1910s and the 1990s – Huisman traces back these heated exchanges to the conversational culture of civic liberalism, the culture of medical learned societies and the liberties of Dutch constitutionalism. In that manner, he resurrects the cultural and political contexts that enabled these experts to embody a specific social role, within which restless audiences invested their trust only conditionally, and often temporarily.

Although very convincing with regards to the expansion of expert authority over ever greater parts of social reality, these conversationalist perspectives may not fully cover the *deployment* of expertise, and the disciplinary power that emerged from the interaction between experts and their objects, rather than between experts and their audiences. What seems to be at stake here, is a sociopolitical divergence of expert paths from the very beginning. On the one hand, the reform programmes of many experts in liberal-capitalist societies (and in their competitors) became a means of expression for yet unarticulated interests and demands, within which sectors of society came to recognize themselves. On the other hand, these programmes simultaneously crowded out other social groups whose interests never materialized in expert discourse, and therefore became subject to objectification rather than being invited into a conversation.

In spheres where public opinion remained relatively absent, expertise certainly was crucial in structuring and reinforcing hierarchies. In her analysis of the Third International Congress on the Scientific Management of Labour in Rome in 1927, Jennifer Karns Alexander discusses expertise as an extension of discipline. By examining six experts who contributed to the congress, she shows how expertise obscured the individuality of workers and was used as a means to ensure required behaviours. A similar analysis of the role of expertise in the exercise of power is conducted by David Niget and Margo De Koster with regard to expertise in Belgian public policy towards 'endangered' childhood. Niget and De Koster examine the scientific practices in the youth observation institutions for juvenile delinquents of Mol and Saint-Servais from the 1910s to the 1950s. They reveal the seemingly contradictory nature of expertise, which seemed to lead to a certain empowerment of the inmates (through the pleas by experts for more responsibilization), but which in the end mainly constrained them (as court judgments could be corroborated by the deterministic approaches of heredity and psychiatry). This contrast, they conclude, constitutes the political rationality of expertise. It reminds us that the historical study of expert performances necessarily entails both the cultivation of particular audiences and the disciplining of particular subjects. If we do not want to merely uphold the stories that modern experts have kept telling about the successes of their performances, we need indeed to analyse the puzzling power of these performances themselves.

1 ETHNICITY, EXPERTISE AND AUTHORITY: THE CASES OF LEWIS HOWARD LATIMER, WILLIAM PREECE AND JOHN TYNDALL

Graeme Gooday

> The best guarantee for the future of the Telephone lies in the extremely modest claims put forward on its behalf by its inventor. Professor [Alexander] Graham Bell – who, by the way, is a Scotchman, and not a Yankee – would probably be the last man in the world to claim for his invention even the possibilities with which it has been connected in the fertile imaginations of American journalists ... With the characteristic penetration of his countrymen, he doubtless understands the difference between a scientific possibility and a practical reality ... [1]

Have experts necessarily been part of dominant cultures? Did their authoritative pronouncements hinge solely on their elite education or on the standing of their powerful patrons? As the volume editors show in their introductory chapter, Eric Ash's account of the early modern 'savant' exemplifies all such features of an 'elite' model of expertise that predates modern cultures of expertise. However, for the period 1860–1960, identified by them as the major phase of state-backed expert cultures, I suggest that these characteristics are neither universal nor permanent characteristics of expertise.[2] To demonstrate that point, I show how three men born into non-privileged families with marginal ethnic-cultural positions in mid-nineteenth-century Anglo-American society all accomplished some form of authoritative status in electrical technoscience by the peak of their careers. Lewis Howard Latimer's birth into the USA's African slave heritage, and the provenance of William Preece or John Tyndall in the United Kingdom's non-hegemonic Celtic cultures did not prevent them acquiring the skills and knowledge needed to perform in key expert roles within their technical institutions.

Such cases show that specialist university training was neither necessary nor actually sufficient for the role of an expert in the courtroom or in other tribunals of Victorian life. As I have argued elsewhere, from the late 1870s to the 1890s the extraordinarily diverse skills required to handle new electrical technologies of lighting, telephony, power and traction could not all be met by one single group or one single mode of training.[3] Accordingly the credentials of electro-technical experts hinged largely on other considerations than any formal education they might or might not have had. For example, no formal education in technology was required for Alexander Graham Bell – émigré Scotsman, Canadian resident and prospective US immigrant – to speak modestly and plausibly as a candidate inventor of the telephone.[4] We see in the above quotation from the London *Times* clear recognition that it was instead his Scottish ethnic pedigree that served as a positive marker of trustworthiness among the technical experts who were pontificating on the telephone's future prospects. On that basis, the partisan *Times* journalist judged Bell to be a more reliable spokesman in prognosticating this technoscience than the inventive American journalists that so avidly followed his telephonic exploits.[5] Clearly, however, the implication was that technical commentators from other ethnic identities might be less immediately trusted than one born in Scotland. In that vein of exploring discriminating judgements about trust, I examine below the contingencies and challenges facing Latimer, Preece and Tyndall in meeting their aspirations. Such challenges were far from trivial in a world for aspirant authority figures who were unconventionally drawn from minority or 'outsider' groups.

The migration of skilled technicians from their homeland – exemplified in the cases of Tyndall and Preece, as much as for Bell – was much more notably a feature of early electrical technoscience than for civil and mechanical engineering. For example, Carl Wilhelm Siemens emigrated from Prussia to England in the 1850s to naturalize as William Siemens; Gisbert Kapp moved from Austria to settle in England in 1881, later becoming the University of Birmingham's first Professor of Electrical Engineering; and the relocation of Irish-Italian entrepreneur-inventor Guglielmo Marconi from Italy to England in 1895. Similarly the early US electrical industry benefited from settling migrants from Europe and North America. In addition to Bell's departure to Canada and later the USA, British electrical specialists migrating across the Atlantic included Elihu Thomson, Arthur Kennelly, Leo Daft and Edward Weston. These were later followed by other major electrical engineers from Europe: Nikola Tesla and Michael Pupin from Serbia and Charles Steinmetz from Germany, all of whom took a major role in raising US electrical technology to international pre-eminence.[6] These figures experienced different levels of success in their transatlantic endeavours: Pupin became a millionaire while Tesla died in poverty.[7] But the chaotic and innovation-hungry world of electrical engineering was generally amenable to expertise of whatever origin, so long as it was economically valuable – whether that economic utility was judged in the laboratory, workshop, factory, patent office or the courts of law. In fact, technologies and expert careers could be financially successful in ways unrelated to ethnic heritage: profit could indeed trump race in the demarcation of who could be trusted as an authority figure.

I thus argue that the diverse ethnic presence in electrical technoscience was not only a manifest fact but a distinctive feature of its enormous fertility in its first quarter-century. Most importantly of all, I show that for the particular cases of Latimer, Preece and Tyndall what they had in common was belonging to institutions that underpinned their expert authority: Latimer to (successively) the Bell, Maxim and Edison companies; Preece to the UK's Post Office, and Tyndall to the Royal Institution in London. In contrast to traditional views of institutions as highly conservative in their orientation, this paper shows that the capacity for electro-technical organizations to assimilate talented outsiders could be part of their symbiotic path to success. To be specific: we will see that this alignment to institutions enabled them to rise to prominence (almost) irrespective of their ethnic identity, far more than if they had sought expert status as freelance individuals. Overall I show that such institutions framed the public nature of expertise enjoyed (in limited ways) by Latimer, Preece and Tyndall: their domain of activity, the kinds of audiences for their expertise, and the duration of that eminent standing. These were just some of the key modalities by which expertise and ethnicity were mutually engaged.

Some Meditations on Authority vs. Expert Status

It is important first to examine the mutable and permeable boundary between 'expert' and authority. Elsewhere I have suggested that in late-nineteenth-century Britain there was a significant, if nevertheless permeable, distinction between them. To offer a brief caricature: an authority was a trusted generalist, able to offer broad, trustworthy – and crucially, disinterested – opinions arising from either broad liberal education or broad experience. By contrast the narrow specialist 'expert' was the antithesis of disinterestedness. As a patronage-dependent servant typically hired to serve the partisan purposes of the paymaster, an expert was not expected to exercise an autonomous judgement or evidential discretion. Accordingly even the most distinguished experts in the profession were subject to the charge of partisanship; hence the elegant insult 'liars, damned liars and expert witnesses' could be directed at even the most prestigious of these.[8] This group of experts has been characterized by Tal Golan as showing an untrustworthy (i.e. ungentlemanly) degree of interest in status or money.[9] Conversely a figure was treated as an authority if they were trusted to be a disinterested oracle or commentator, not paid for their testimony or subject to the controlling interest of any other. This 'authority' was typically an individual asked to comment from their capacity as a president of a learned society or a professor at an academic institution.

We see here a key theme highlighted by Vandendriessche, Peeters and Wils: that whatever it meant to be an expert or an authority, it was necessarily with respect to a specific audience, and with a particular kind of performance-based relationship to that audience. After all, what could it have meant to be an 'authority' whose views nobody took seriously, or an expert in a topic that nobody wanted to know anything about? We thus see the need to eschew reductive treatments of expertise and authority as if they were static matters of social class or education, but treat them rather as a dynamic disposition that rested on the contingencies of audience evaluation of performance – not on the philosopher's abstract idealization of epistemic omniscience or at least infallibility.[10]

While in this chapter I do not suggest that we abandon the distinction between expert and authority, I argue that the boundary between these categories became blurred by the early twentieth century. Being an 'expert' lost its pejorative status as the economic reliance on paymasters decreased, and successful freelance experts were free to comment without heavily constraining obligations to patrons – albeit well remunerated by the legal system. It was in this way that Lewis Howard Latimer came to be recognized and respected as an 'expert' beyond his employment by several leading electrical engineering companies in the USA that effectively competed for his services.

Concurrently, as Vandendriessche, Peeters and Wils remind us, the issue of specialization also came to matter. The status of individuals to comment from positions of authority within institutions lessened as more specialist knowledge became required to command the respect of an audience. Being a specialist, rather than the broadly knowledgeable generalist, became increasingly the model of proper learning for public consumption. This, we shall see in the cases of Tyndall and Preece led to their reputations fading after their death. Although they acted during their careers as institutionally grounded authorities on a broad range of topics, at the end of their careers the sheer breadth of their expertise became subordinate and less valued than the more specialized knowledge of the 'expert'. Conversely, Latimer's reputation grew from being very narrowly focused on the technicalities of legal-engineering advice, to becoming a true polymath. Hence now in the twenty-first century we find that the roles of authority figures and experts are much less distinguishable than they arguably were in the nineteenth century.

Another kind of point that I wish to make is that non-dominant ethnicity was less of a barrier to public expert witnessing or authoritative pronouncements in science than either gender or disability. While males from non-dominant groups could be experts and authorities in a range of key ways, neither women nor the disabled adopted such a public role in the late nineteenth century. Early 'expert witnesses' in court were invariably (and perhaps unsurprisingly) typically male and able-bodied. As I have argued elsewhere, female authority was so often limited to the home, and this almost invariably limited their activities to writing advisory manuals or journal articles for other women; presumptively they had the authority to write on the organization of the home in ways that their menfolk did not. But it is very hard to find any evidence of women appearing as an expert witness in a court case, or lecturing on electricity, until the twentieth century (specifically after the First World War).[11]

It is notable correlatively, that those with disabilities such as partial or complete hearing loss, did not generally take part in the kinds of performative role that rendered them as public 'experts'. In the world of early electrical telecommunications, a significant number of leading individuals were partially deaf – such as Oliver Heaviside, Thomas Edison, Alexander Muirhead and John Ambrose Fleming – and did not participate in cross-examinations or public debates for obvious reasons. As for court judges who were deaf – that was another matter. While Edison and Fleming both appeared occasionally in public, they tended not to get in involved in dialogic discussions, typically spending more time in the workshop/laboratory inventing, or in their studies writing. Neither Heaviside nor Muirhead ever appeared in public, their severe deafness and reticence meaning they instead wrote letters intensively to communicate. So insofar as Heaviside was any kind of recognized expert or authority figure in electrical communications it was because his incorrigibly dense mathematical writings impressed a few senior figures in the profession such as the Maxwellians and Lord Kelvin. As Heaviside's biographers melancholically note, he only really became immortalized among engineers as the iconic electrical researcher after his death in 1925 – that is once the cessation of his misanthropic and prickly missives gave his followers scope to establish memorials to him.[12]

The converse was the case for Preece and Tyndall (and to some extent even Latimer) whose performance-based reputations faded after their respective deaths. But this point itself is revealing: much of their role as expert was from their *performative* capacity with the spoken word – whether under cross-examination in the witness box, or as authoritative figures in the lecture hall. As we shall see, Latimer was a very dextrous respondent to cross-examination in the courtroom, and both Preece and Tyndall were (for the most part) superlative public speakers: it was their forceful verbal performances, repeated regularly through the cities in which they spoke, that rendered them authority figures. From this we can see that the able-bodied capacity to deal speedily with real-time dialogue was what the 'hard of hearing' found most challenging, and thus why the partially deafened characters described above were hardly ever to be found in the British courtroom. Conversely, once the expert performances ended, so too did expert status.

At the same time, what about the ethnic markers? Whilst Latimer's African-American slave heritage was not something he sought to deny, it is revealing that both Tyndall and Preece generally kept their distinctive regional/national accents in check whilst in public performance – unlike Scotch-born migrants like Bell, who seems never to have sought to hide the linguistic markers of his ethnic origins. But this was not the case for others with Irish family heritage. The controversial innovator in early wireless telegraphy, Guglielmo Marconi, cultivated assiduously unaccented fluency in both Italian and English (from his Protestant Irish mother Annie Jameson) in order to move smoothly through his two native lands without being dismissed as a 'foreigner'. Tyndall was ultimately less successful at this, however. As we shall see, it was at Tyndall's final failing lecture at the Royal Institution in 1886 that at time of collapsing self-control his long-suppressed Irish brogue returned intermittently amid his carefully Anglicized diction.

From this we can return to the point about the audience-constituted nature of expert or authoritative standing raised by Vandendriessche, Peeters and Wils in their introduction. They emphasize that the socio-political 'empowerment of expert performances' by specific audiences was intimately connected with the effectiveness of a particular expert encounter. We shall see that even with years of learning, experience, institutional support and public popularity, the status of expert could be undone by performances that were not effective in maintaining an authority-relationship with their audiences. As the editors emphasize, the quality and credibility of such performances were of course judged by audiences – lapsing into vernacular or some non-standard version of the mainstream language could betoken a lapse from authority into questionable marginal status. To some significant extent then we need to understand credibility as something that has been *attributed* by an audience – whether explicitly acknowledged or not – in relation to the expert's successful performance (or otherwise) of key verbal markers. To reiterate the point: status as an expert or authority could be enabled by either education or institutional standing, but not thereby guaranteed.[13]

Lewis Howard Latimer: The Expert Rising from Slavery

As several ethnically-sensitive revisionist US historians have recently shown us, there were some African-American contemporaries of Edison, Preece and Tyndall who acted as a vigorous, but barely remembered, constituency of inventors and consultants in the innovative technologies of the late nineteenth century. In North America the outstanding example in electrical technoscience was Lewis Howard Latimer, who was the first son of a former African-American slave to become a public expert. Having worked for Bell in the crucial telephone patent applications of 1876 (allegedly even drafting the final version of Bell's specification), Latimer worked for Hiram Maxim in the USA before moving briefly to work for the UK-based Maxim Weston company in England. Here, however, he was clearly not accepted by the management since he declined to show due deference to his employers, and this was influential in his departure in 1882.[14]

After returning to the USA, Latimer later recalled the great challenge he faced: 'Here we found the ranks closed and every place filled'. After a few months of casual work, however, 'the Edison People sent for me and I became one of the firm'. In this version of his autobiography Latimer emphasizes his smooth rise up through the ranks to become engineering department draughtsman, thence to the Edison (later General Electric) Company's legal department in 1889. While he moved on to other employment in 1911, he was remembered by Edison and others in that company as 'one of the pioneers in the electric lighting industry from its creation until it became worldwide in its influence'.[15]

And yet, from another autobiographical source, presumably written for a different audience, we know that Latimer did at first encounter racist responses to the unusual sight of an African American in a technical drawing office. As his autobiographical 1911 *Logbook* reveals (writing in the third person of himself), it was the economic value of his high skill-level that won over the sceptics:

> Now his color began to be a draw back to him. Every new workman who came into the office saw for the first time, a colored man making drawings; and as often as they came to work in the office they tried to pretend that he could not do their work. But he had had such long experience and was so well posted in all kinds of drawing that they soon were forced to acknowledge his exceeding ability which was far above the average at that time.[16]

It was not only through his skilful demonstration of both drawing and drafting of patent specification text that he proved to be indispensable to Edison. He soon served as a patent expert in the numerous legal challenges to Edison's patents made by rivals.[17] According to William J. Hammer, his friend and fellow Edison-hired expert, Latimer 'made drawings for court exhibits, had charge of the library, inspected infringing patents in various parts of the country, testified as to facts in a certain number of [court] cases'. Indeed it would appear that Latimer became the expert of choice. For example, in 1891 at a meeting of Edison's Patent Litigation Committee 'our Mr Latimer' was reported as a key individual invited to give testimony for the Western Electric Company to break down a rival's patent.[18] Eventually Latimer's success as an expert witness was so great in cases of patent litigation that he ended up being chief consultant for the US Board of Patent Control formed by General Electric and Westinghouse to pre-empt the destructive occurrence of mutual patent litigation. In what appears to be the draft of an obituary note, one colleague recalls:

> Latimer was always a valued aid to Thomas Edison. Edison was repeatedly involved in lawsuits which he had to institute against the companies who infringed upon his patent. In all of these suits, some of which involved millions of dollars, Latimer as the original draftsman was Edison's star witness, and the suits were often decided by his testimony.[19]

Overall, Latimer's success in electrical technoscience was a major accomplishment by most standards, and all the more so to arrive at a position of great trust for the son of a former slave with barely any formal education. After his retirement, as further recognition of his role the Edison company paid its debts to Latimer by enrolling him into the league of Edison Pioneers – albeit as the only African-American member.[20] Nevertheless, as Rayvon Fouché notes, Latimer's strategy of assimilation into the mainstream white professional elite was not without cost. Throughout his ever more successful professional life, Latimer's public utterances passed little comment on the controversial contemporary politics of race. For his capitulation to the former slave-owning class, more radical groupings who sought to promote distinct African-American priorities and values thus categorized Latimer as a 'Negrosaxon'. For them, being an authority did not include performing as a hired expert for an American capitalist corporation.[21]

William Preece: Chief Electrician at the UK Post Office

In late nineteenth-century Britain, unlike in the USA, patent litigation was more the province of highly educated individuals with the verbal dexterity to outmanoeuvre rivals in highly ritualized courtroom disputes.[22] Instead the Welshman William Preece found a natural home in the General Post Office (GPO) telegraph service and the Irishman John Tyndall in the Royal Institution. In such places they could gain access to resources for research and direct a considerable influence on the trajectory of technoscience in their institutions. Neither chose to be expert witnesses in court cases – and Tyndall even refused to have anything to do with patents. Accordingly both might have hoped to be regarded as relatively disinterested spokespersons on the subject using their institutional bases from which to offer trustworthy pronouncements. Preece and Tyndall were more comfortable in the roles of public authorities *qua* educational lecturers, testifying for government commissions on pressing points of electrical light and telecommunication, not directly serving as a spokesman for capitalist interests.[23]

Preece's career began with an early departure from Wales to accompany his bankrupted father to London for a fresh start in life. Although more privileged initially than Latimer, Preece's higher education at University College London was abandoned (more time being spent on cricket than advanced study). Instead Preece like Latimer started at the bottom of the professional hierarchy, gradually ascending from General Post Office (GPO) apprentice in the late 1850s, to become its chief electrician forty years later.[24] As I have shown elsewhere, Preece used his position in the GPO to fashion for himself a role as *the* British national authority on the nature and applications of electricity throughout the 1880s. For a nation puzzled by and somewhat fearful of the strange and dangerous powers of electric lighting and power revealed by the operations of Edison, Swan, Hopkinson et al., Preece's lectures on electricity served to reassure many in the public sphere that at least *somebody* knew what electricity was and what it could do.[25] Building on such an authoritative reputation, he was able to serve lucratively as a consultant on many heavy-duty electrical engineering projects (especially the municipal variety) without himself ever having been trained beyond the management of telegraph wires.[26] At no time was there any report that his discursive manner was other than standard English: as an authority figure for the British Isles based in London, he adopted the modes and mannerisms of the English gentleman.[27]

As Elizabeth Bruton has shown, Preece's career at the General Post Office did, however, engage geographically with his home country Wales more often than others in his profession – never estranging himself from his roots as both Latimer and Tyndall were to do. The GPO in London was an active site of innovation in relation to the 'crossed-wire' interactions of telegraphy and telephony in the 1880s that led to an early system of 'inductive' wireless signalling being developed for the Post Office in the early 1890s. Preece and his staff ran trials of the system across the Bristol Channel to test its capacity (amongst other things) to serve as an alternative to submarine cables if they ever broke down. That part of Wales was by then one of the few parts of the UK still far enough away from sources of electrical interference, including telegraph cables, telephone wires, trams, electrical lighting and so forth, to get reliable results. Although not very adaptable beyond such short-range scenarios, this inductive wireless system was arguably enough to attract the young Marconi to Britain in 1895–6, seeking the patronage of Preece at the Post Office, to get support for his new 'signalling' version of Hertzian wave apparatus.[28] Twice in his career Preece's authority in Britain extended beyond the state-run Post Office to become President of the Society of Telegraph Engineers and Electricians (STEE, renamed the Institution of Electrical Engineers from 1888). He did not hesitate to use his authority in this position to quell rival views to his own, not least those of Oliver Heaviside. As is well known, Preece was his nemesis. In his publications on mathematical telecommunications theory between 1885 and 1892, Heaviside cast himself as a public-minded altruist, giving his knowledge freely to the world. This was in contrast to Preece's vigorous public persona, and in particular his use of his GPO status to suppress Heaviside's inconvenient challenge to Preece's view that long-distance telephony was closely analogous to telegraphy. At a meeting of the STEE in January 1887, Preece refused to accept Maxwellian views (as adopted by Heaviside) that telephone lines, unlike submarine telegraph cables, needed enhanced self-induction to transmit speech at long distance. Heaviside's joint paper with his brother Arthur was soon rejected by the STEE and several other key electrical journals: some inferred that Preece had used his powerful position to censor Heaviside's heretical Maxwellian claims.[29] However, once Sir William Thomson announced his approval of Heaviside's self-induction-based approach to optimizing long-distance telephony, *The Electrician* once again published Heaviside's work. Preece's position was thus defeated by Heaviside's alliance with a figure of greater authority than Preece – Sir William Thomson, soon to be Lord Kelvin. Nevertheless, Heaviside's loathing of Preece's behaviour prompted him to develop a broad range of insulting epithets, such as the sardonic phrase: 'Eminent Scienticulist'. Tellingly, in letters to friends Heaviside alluded to Preece's origins, derisively dubbing him 'Taffy' – a derogatory reference to his Welsh identity.[30]

Preece himself only publicly acknowledged his Welsh identity once he had retired from professional life in 1899 and moved back to his home town of Carnarvon. There he was chiefly instrumental in setting up the Pan-Celtic congress, publicly linking his own Celtic heritage in Wales to that of the Irish, Scots, Bretons, Cornish and Manx through explorations of literature, poetry, song and shared mythologies. It is easy to see this post-retirement move to intense participation in Celtic culture as an indication of Preece's voluntarily repressed ethnic identity throughout his professional life – a condition of success to which he no longer needed to subscribe after departing from the constraints of his public role at the Post Office. Nevertheless, after Lord Kelvin died in 1907, as his longest surviving colleague Preece was invited by Lady Kelvin to speak at the Institution of Electrical Engineers commemorations in London.[31] So even as the GPO declined to maintain Preece in high standing, others with whom Preece had disagreed still deferred to him in later years.

Let us turn finally, to the case of another ethnic Celtic – a near contemporary of Preece – whose rise to fame through naturalization in an 'English' gentlemanly culture ended somewhat differently.

John Tyndall at the Royal Institution: The Skilled Irish Orator

As Jill Howard has cogently explained, John Tyndall had a challenging early career. He rose from humble origins in southern Ireland (from a Protestant family in a predominantly Catholic culture) via work as a surveyor, moving then to Germany to secure an education in the physical sciences. By the late 1850s he was securely established as professor of natural philosophy at the Royal Institution (RI) in London, initially alongside Michael Faraday, the professor of chemistry. There Tyndall secured enormous popularity as a flamboyant and very efficacious lecturer.[32] This popularity was reflected in the sales – tens of thousands globally – of published versions of his lectures. Thus *Fragments of Science* (1870), *Six Lectures on Light* (1873) and *Lessons in Electricity at the Royal Institution* (1876) helped to make him one of the wealthiest and most sought after of British scientific men in the 1870s. Much of this was achieved through Tyndall's very popular performances at sold-out lecture tours in London and beyond, many (though not all) focusing on electrical topics.

It was in such an extra-mural setting that Tyndall spoke more freely, as made famous by his controversial Belfast address to the British Association for the Advancement of Science in 1874, interpreted by many in the Church of England as a manifesto of agnostic materialism.[33] That being said, his reputation as a popular speaker outside the RI, perhaps one of the best known Irishmen of his day, led others to seek him out as a crucial agent of calm when the population of Ireland rebelled against English colonial occupiers. Tyndall was asked later that same year to go to Dublin to give a series of lectures on science to Irish working men to give them non-political matters to consider as an alternative to outright revolt and insurrection. As the secretary of the Royal Institution, William Spottiswoode, wrote to Tyndall later that same year:

> I can hardly say how earnestly I desire that you should do this bit of work for us. Your connexion with Ireland, your sympathy with the working men (strong, & mutual), & other considerations which you well know without my recounting them, all point to you as the man. It is a great thing to ask of you, I know, but this is an occasion which happens to me only once in life; & it is one on which I need all the support which my best friends can give.[34]

Nevertheless, Tyndall declined to go back to the land of his birth for fear of controversy. As he replied to Francis Galton in view of the aftermath of the Belfast address: 'the Archbishops, Bishops & Clergy of Ireland have banished me from their soil with bell, book & candle-light, as St. Patrick banished the snakes & toads. How I am to [fare] among the working men of Dublin is easier imagined than described'.[35] Tyndall's presumptive audiences were obviously not the Irish readers of his works.

Instead, following his very successful US tour in 1872, in the last decade of his career (1876–86) Tyndall preferred to spend time either in England or on therapeutic alpine treks. He made an enormous impression not only on the popular lecture audiences of London, but also on a rising generation of physicists. As Oliver Lodge said in his autobiography, Professor Tyndall became 'one of his heroes'. After his first trip in 1866, Lodge attended every Friday evening, when Tyndall was 'holding forth at the Royal Institution', and would do anything he could to get a ticket.[36]

Silvanus Thompson, later a close friend and lifelong correspondent of Lodge, had a similarly inspirational first encounter with Tyndall ten years later than Lodge in 1876, albeit laced with some criticism. Thompson had been studying with Frederick Guthrie at the Science Schools (now part of Imperial College) in South Kensington. After studying electricity with Guthrie, Thompson found that Tyndall's approach was so much more interesting. In a letter to his mother of 6 January 1876 Thompson wrote there was 'a dash and an ease about Tyndall's speaking and manipulating' not seen in Guthrie's lectures.[37] Thompson reproduced something of the mode of Tyndall's lectures to children when he took up his post at University College, Bristol in 1876, offering a Christmas holiday course to juveniles on 'Voltaic Electricity' in 1877–8. Tyndall wrote to Thompson 'I wish you success. Your movement, depend on it, is an important one'.[38]

Ironically Thompson's widow narrated in *Silvanus Phillips Thompson: His Life and Letters* (1920) that the student Thompson soon had another new oratorical hero, T. H. Huxley, whom he heard lecturing 'Upon the Comparative

Anatomy of the Lower Vertebrata' in spring 1876. Thompson related to his mother: 'the flow of language was perfect and the whole manner most graphic, perspicuous, and simple'. As a speaker he 'beats Tyndall hollow'.[39] This is noteworthy since Mrs Thompson (originally Jane Henderson) – herself originally a reporter for the *Glasgow Daily Mail* – was an astute critic of lecturers and first encountered Silvanus delivering a lecture at the Quaker Friends of Westminster meeting later that same year. Having already heard Huxley speaking, on encountering the young Silvanus 'she much appreciated the powers of the young lecturer, and especially admired the beautiful peroration with which he closed'.[40]

She had not yet, however, heard Tyndall speak and did not do so until 22 January 1886. And when she did so, she was shocked at what appears to have been Tyndall's career-terminating lapse into dependency-induced performance failure. At the Royal Institution venue in which he had charmed and authoritatively educated many in the three preceding decades on his key discoveries in atmospheric science, magnetism and more besides, Tyndall's gravitas now finally eluded him. As Mrs Thompson reported:

> I was looking forward with pleasure to this lecture, which was to be given by Professor Tyndall, who was so famous as a scientific expositor. The subject was 'Wave Forms'. But such a disappointment! Poor old man, he maundered on for an hour and twenty minutes, repeating himself over and over. He gave us a life of Thomas Young with very little mention of 'Wave Forms' at all. He is a weird-looking, thin, stooping old man with long grey hair hanging from a high, narrow head. He has a decided Irish brogue now and then, and uses curious gestures.

Once its 'brilliant resident professor', he now revealed a long-suppressed Irish vernacular that could not be hidden in such less self-managed episodes.[41]

Revealingly, this is the only report among his contemporaries of Tyndall's verbal delivery in the broad Irish diction of his youth, and suggests perhaps just how much work he had put into masking his ethnic origins throughout so much of his career.

After this fiasco, Tyndall never lectured at the Royal Institution again, and from March 25 1886 Tyndall's course of lectures on light was swiftly replaced by James Dewar's course of four lectures on electro-chemistry. [42] There were brief hopes of a return in November that year following some time spent recuperating in Switzerland, but by April 1887 he had resigned from the RI.[43] When Tyndall died six years later, the obituarists diplomatically made no reference to this final failed lecture, but recalled him instead at the height of his powers. Silvanus Thompson in particular argued that such was Tyndall's great skill as a lecturer that his more controversial activities and utterances could – and should – charitably be overlooked:

> Possessed of a fluent and impressive delivery, rising at times into real oratorical power, his lectures attained a charm of style unique and inimitable... He made the whole

world his debtor by his masterly expositions of modern science, and the whole world may well forgive the exuberances of his ardent nature. [44]

It was through this reference to Tyndall's 'ardent nature' that Thompson alluded to what was then construed to be Tyndall's 'Celtic temperament' – and perhaps also to the consequences of apparent alcoholism and drug addiction that seem to have played a major role in the diminution of Tyndall's career in later years. Indeed, very soon his reputation as a specialist professional who had revealed why skies were blue, and how atmospheres rich in certain gases were amenable to net absorption of heat (global warming), was forgotten. Instead his posthumous reputation, narrowly construed as that of a generalist lecturer, was under subtle attack even among his own devotees. As Lodge wrote of the late Tyndall in the *Encyclopaedia Britannica* of 1902–3:

> His knowledge of physics was picturesque and vivid rather than thorough and exact, and never did it make any pretence at being encyclopaedic. In amount it did not much exceed what any highly educated man of genuine all-round culture should aim at, though it was far in excess of what in England is commonly thought possible; but its vividness and colour and garnish of enthusiasm were his own. His strong, picturesque mode of seizing and expressing things gave him an immense living influence both in speech and writing, and disseminated a popular knowledge of elementary physics such as had not previously existed...[45]

In fact, Lodge's view was not now that Tyndall was an authority in matters of science per se. Tyndall's status in the physical sciences barely recovered from such attack – at least not until twenty-first century revisionism returned to evaluate the global significance of his research on the differential effects of atmospheric warming.

Conclusion

We have seen that to become an authority figure in late-nineteenth-century electricity, neither a higher education nor mainstream ethnic identity were strictly required. A strong sense of affiliation to an organization with patronage was, however, clearly necessary: Latimer with the US Edison company, Preece with the UK's General Post Office and Tyndall with the Royal Institution in London. Having built their expertise and relationships with their audiences in those domains, each maintained their expert standing while they were still affiliated. There was more to this than mere institutional affiliation, however – such an analysis could apply to the career trajectory of many others without any visible ethnic marker. For the three figures studied here, there was the matter of how they managed the cultural-ethnic differences that any informed observer might use – and at least occasionally did use – to undermine their authority.

Notwithstanding their unprivileged origins, and unpleasant experiences earlier in their careers, these three diverse examples of Anglo-American experts/authorities clearly succeeded in securing at least some level of major audience recognition for their expertise among mainstream audiences. This was most striking for Latimer, whose African heritage status was perceptible to all who saw him in performance. By contrast, for Preece and Tyndall, their verbal performances generally disguised audible markers of cultural difference with some care so as to maintain a strong relationship with their audiences. At the height of their careers, all three were highly skilled in expert performance – delivering what was needed and in a theatrically effective fashion. However, once the living memory of performance was gone, their status as authority was diminished: hence Latimer, Tyndall and Preece were little remembered a generation after their deaths. As a rare example of African-American professional eminence in electrical technoscience, Latimer's reputation thrived only among his assimilationist sympathizers – although recent historical research has rehabilitated his reputation somewhat. While Preece has recently became a national hero of Welsh science, Tyndall was an ambivalently cherished figure in Ireland, and now famed more for his discoveries concerning the mechanisms of blue skies and global warming, with little attention drawn either to his agnosticism or later decline in standing.

That being said, we can return to one key difference in career trajectory that was highlighted by Vandendriessche, Peeters and Wils in their introduction. This is the matter of specialism vs. generalism and the increasingly pre-eminent role of the former during the period 1860–1960. This worked most effectively for Latimer. In the US culture of expertise Latimer premised his reputation on ever more narrowly specialist knowledge in electrical technology. By contrast, Tyndall and Preece both became ever more generalist in their advancing years: while such breadth was a prerequisite to epistemic authority in their mid-nineteenth century youth, by the turn of the twentieth century such was no longer the route to advanced standing in the technosciences, with all their complex division of labour. Thus we can see why, whereas Latimer built up a loyal institutional base in Edison's team as a virtuosic specialist that lasted after he retired and unto death, Tyndall's and Preece's retirements saw them lose almost all of their connection with their former employers' institutions, and in Tyndall's case much of their public prestige. Thus we see that the double strategy of institutional loyalty and an expedient move to specialist knowledge was the key to Latimer's authority within mainstream American culture.

Overall, however, given such contingency and locality we can see that what it is to be an expert/authority was not self-evident in late nineteenth-century Anglo-American culture. It was not just a matter of social class, heritage or education: rather it was a matter of trustworthy performance. Most importantly the path to becoming an expert involved getting the patronage of a strong institution, choosing circumscribed domains in which to be an expert, and constructing positive relationships with audiences in those domains. With technical skill assured and economic benefits demonstrated, such socio-political aspects of being an expert/authority figure could overcome some of the deeper boundaries of arbitrarily contrived cultural-ethnic 'difference'. Accordingly, future historians can usefully interrogate how a performance-based understanding of the subject can illuminate historical pathways to expertise in ways that transcend conventional assumptions about the significance of personal heritage or cultural identity.[46]

2 ARBITERS OF SCIENCE: EXPERTISE IN PUBLIC HEALTH IN NINTEENTH-CENTURY BELGIAN MEDICAL SOCIETIES

Joris Vandendriessche

In 1864, an extraordinary meeting was organized by the Medical Society of Ghent, a club of Belgian physicians of whom most were affiliated with the city's university. The reason for the meeting was the latest study on the living conditions of Ghent's worker population by the physician and professor, Adolphe Burggraeve. In a plenary speech, Burggraeve reflected on his own motives and, more generally, on the social role of the physician. 'It does not suffice for the physician', Burggraeve addressed his medical colleagues, 'to engage in the medical sciences to keep abreast of their progress ... it is necessary in addition to devote oneself to the improvement of the hygienic conditions of suffering humanity'.[1] Burggraeve himself seems to have lived to his own ideals. After being appointed professor of anatomy and surgery in 1830, he published a series of scientific studies on surgical instruments and techniques.[2] But what garnered him even greater fame was his work within the field of public health. As a socially and politically engaged physician – Burggraeve resided in the town council of Ghent in the late 1850s and throughout the 1860s – he spread his views on sanitation, the prevention of epidemics and personal hygiene in numerous treatises, plans, studies and booklets intended for an equally diverse public of physicians, politicians and laymen.[3] By the middle of the century, Burggraeve was an established expert in public health.

By the turn of the twentieth century, however, university professors such as Burggraeve had become a rare breed among Belgian public health experts. Most experts in the field of public health were now employed full-time by the state in the growing number of health services and commissions at either the local or the national level. School doctors, health and food inspectors, bacteriologists, statisticians and many others formed a new professional group of physicians working in civil service.[4] With their emergence, a tradition of part-time and voluntary

'philanthropic expertise', of which academics such as Burggraeve were representatives, seemed to have faded away. Instead, a new generation of public health
professionals presented themselves as scientific specialists. The foundation of the
Société Royale de Médecine Publique in 1876 illustrates their institutionalization in the late nineteenth century. In the early twentieth century, professional
training followed as specialist education was being offered at the university.[5]

In the historiography of public health, this emerging group of public health
professionals has attracted far more interest than 'philanthropic' experts such as
Burggraeve. Public health, as a field of expertise, has been studied in the first place
by looking at the interaction between these professionals and state officials.[6] Following a thesis first presented by Oliver MacDonagh, expertise in public health
developed parallel to the gradual expansion of the state, which ever more appealed
to experts for policy making and reinforced in this way the professionalization of
advisory functions.[7] More recent work has pointed to the growing embeddedness
of this new group of professional experts in contemporary ideological debates.
As matters of public health became increasingly politicized in the course of the
nineteenth century, experts also became ever more part of politics.[8] The health
policies of the state have therefore been the primary lens through which the development of public health, as a field of expertise, has been examined.

This chapter aims to broaden our view on the emergence of public health
expertise by using a different perspective. Instead of focusing on the budding
group of public health professionals who operated in civil service, expertise in
public health will be studied through the eyes of medical academics. Such a
perspective allows, first of all, to present the vanishing of a tradition of philanthropic expertise as part of a larger reorientation of the field of public health
in the second half of the century. Both groups – academics and public health
professionals – had to reposition themselves in a medical landscape in which the
boundaries between scientific, professional and popularizing efforts were drawn
ever more sharply. In this transition, academics continued to be active players
by taking up new roles of arbitration, judging which expert studies could be
regarded scientific. Second, such a perspective allows for examining the extent to
which these new professional experts evolved out of established traditions of scientific advice. Attention will not only be paid to the recasting of existing expert
performances and the creation of new ones, but also to what was lost when the
embeddedness of expertise in a tradition of philanthropy was left behind. Some
of the problems of credibility and authority, which came along with the salaried
positions of experts in civil service, can be considered the result of this transition.

Such a perspective on expertise can be realized by looking at a specific performative setting: the urban medical society.[9] In these societies in the major
Belgian cities, such as the one in Ghent where Burggraeve presented his study, the
academic community was the leading medical group. They were places where the

tradition of philanthropic expertise flourished in the middle of the nineteenth century through speeches, reviews and discussions during the society meetings. From the middle of the century, state-employed public health professionals tried to become members of these by then prestigious societies. In this chapter, the interaction between both groups will be studied by looking at the medical societies of Brussels and Ghent. Attention will be paid to the professional trajectories of their members, the studies that were presented and the audiences these studies aimed at. The emphasis will be put on the period between 1860 and 1880, a turning point in the evolution of public health expertise in Belgium.

After a sketch of the tradition of philanthropic expertise in urban medical societies, the professional development of public health specialists will be briefly discussed in relation to the changing hierarchies in the field of medicine. Next, the inclusion of these new experts in medical societies will be examined. Their participation brought along new scientific practices and expert performances. In the final parts of this chapter, the debates on political engagement and the public role of the medical sciences between public health professionals and medical academics are analysed. These debates are telling of some of the problems that arose together with the professionalization of public health expertise.

Philanthropic Expertise

At the time of Burggraeve's speech, urban medical societies had developed into well-established institutions in the Belgian medical landscape. In the major cities, they had been founded in the 1820s and 1830s. In Brussels, the Société des Sciences Médicales et Naturelles de Bruxelles was founded in 1822 and in Ghent, the Société de Médecine de Gand was established in 1834. Upon their foundation, these urban medical societies represented a relatively new organizational form because of their private nature, exclusively medical membership and focus on scientific research. These features made them different from the more established, semi-public learned societies, such as the Académie des Sciences, des Arts et de Belles-Lettres.[10] The members of these new societies consisted of different medical groups: academics who taught at the university, hospital physicians, private practitioners, military physicians and pharmacists. University professors were the leading group, taking up the positions as board members, although they did not form the largest professional group. In the Medical Society of Ghent in 1840, for example, only seven of the thirty-two members of the society were university professors. The other members comprised of six physicians working in the city's hospitals, twelve who held a private practice, three military physicians and four pharmacists.[11] In Brussels, the group of private practitioners was even more substantial. Of the thirty-one members in 1840, eighteen were private practitioners, seven were academics, five were military physicians and two were pharmacists.[12] What united these different groups of physicians was their approach towards medical science. Medical observations were discussed and studies were reviewed during society meetings. The results of these efforts were published in the journals of medical societies in the form of meetings reports, individual studies and reviews. These publications testify to a diverse scientific community in the middle of the nineteenth century.

It was the leading members of such communities that took on expert roles in public health. Mostly academics and established military physicians were appointed in the emerging public health institutions of the Belgian government. These institutions consisted in the first place of local and provincial health committees, but were supplemented in the 1840s by new central institutions such as the Académie de Médecine, which was founded in 1841 with the double goal of advising the government and advancing science.[13] A typical example of a successful expert within this world of emerging state commissions was the professor of obstetrics in Ghent, Alexis-César Lados. In the 1830s, Lados had rapidly built an academic and scientific career in obstetrics and legal medicine as a member of various medical and scientific societies, including the Medical Society of Ghent in 1837. 'His remarkable works', as his biographer later noted, 'attracted the attention of the government'.[14] In 1841, Lados was indeed appointed a member of the provincial medical committee of East Flanders, of which he would later, in 1865, become president. In 1862, he had already been elected an honorary member of the Académie de Médecine.[15] These part-time advisory positions in government service were regarded an addition, a mark of appreciation almost, to a wide range of public and private functions.

On the institutional level, the divisions between private and public institutions were equally fluid. Medical societies, in addition to government committees, could also provide expert advice. What made these societies so attractive for government administrators was their impressive scientific production and quickly rising scientific reputation. More than traditional learned bodies, urban societies realized a considerable body of scientific scholarship. In 1851, the members of the Medical Society of Ghent could for example proudly declare, presenting their society to the urban government, to have published twenty-seven volumes of their *Annales* and eighteen volumes of their *Bulletin*.[16] For the government, involving these prestigious scientific societies was a way of legitimizing the social and public health policy, although limited and slowly emerging, of the relatively new state of Belgium to the (voting) urban bourgeoisie.[17] The form of cooperation which emerged in mid-nineteenth-century Belgium, as elsewhere in Europe, was one of the outsourcing of social issues to the private scientific community. Such collaboration was financed by subsidizing prize questions and specific studies.

The most well-known example of such expert studies on demand was the inquiry into the working and living conditions of the workers in Ghent's cotton mills in 1845. The study was requested by the Belgian Ministry of Internal Affairs and conducted by the physician and professor of chemistry at the University of Ghent, Daniel Mareska, together with his colleague, the physician Jean-Julien Heyman.[18] Its presentation at the Medical Society of Ghent, as the meeting report indicated, was 'covert with unanimous applause' by the members present, who decided to publish the study in their journal.[19] Because of its clear description and analysis of the working class conditions, it became an often-cited work, both in medical and governmental circles. In Belgian social history, the study has therefore been said to form a crucial stepping stone in the development of social policy. It has also often been used as a testimony of the poor living conditions that resulted from the industrial revolution.[20] Such collaboration between the state and medical societies continued well into the 1860s. In 1861, the same ministry allocated a subsidy of 300 francs[21] to the Medical Society of Ghent to finance a study – in the form of a prize question – on the hygienic measures to be taken in schools.[22] The next year, the urban government sponsored a similar competition on 'the diseases that may have their origin in the linen and cotton industries'.[23] Advisory roles were thus common practice in urban medical societies.

Such interaction with the government meant more than merely financial support. The fact that state officials turned to these institutions also entailed a legitimization of the scientific knowledge of their leading members.[24] It allowed medical academics, more concretely, to claim a general interest, a social importance for their scientific work. It is within this context that experts' rhetoric of philanthropy and social duty, which was also present in Burggraeve's speech, needs to be understood. Taking up an advisory role as public health expert was a way to increase one's status and reputation. The initiative, however, did not always come from the government. Socially engaged expert studies of health developed into an independent genre within the medical literature. Professors in Brussels as well, such as Jean Dieudonné, published their medical views on what came to be known as 'the social question'.[25] Within such a context, government subsidies were never perceived as a salary, but rather as a normal support of scientific research. When the Ghent society members discussed for instance how to warn the public of the dangers of certain matches causing injuries and fires, Lados called for an investigation into the alternatives for phosphor ones. Such a study, so he addressed his fellow members, deserved 'a good reward' by the government.[26] By the 1860s, it seems, many medical scientists expected, even demanded, the state-support of socially redeemable medical inquiry and practice.

The success of this system of reciprocal legitimization depended on its ability to reach a wide audience. Not only government officials, of which some became honorary members,[27] needed to reached, but also their fellow society members and more generally the readers of the societies' journals. Such a wide audience explains the theatricality that seemed inherent to the presentation of expert studies in medical societies. Social engagement of experts and their institutions, and government satisfaction and continued demand for expertise, needed to be emphasized. Both explicit philanthropic ideals of 'warning the general public' or 'alleviating the suffering of humanity' and the appreciation of government officials were important in this regard. In 1860, for example, when the members of the Medical Society of Ghent met the royal family during a visit to the city, the Belgian king referred, in an answer to a speech by Lados, to the publications of the society and the poor hygienic conditions of the working class, encouraging them in their work: 'Continue, gentleman, together with my government to work for the improvement of sanitary condition of this region'.[28] Such a speech, rhetorical in its meaning as it was and published in the society's journal, could nonetheless convince both medical practitioners and state officials of the social relevance of the society. An even broader audience could be reached when studies were picked up in the general press. In 1863, the editorial board of the journal *Le Progrès* asked the approval of the Ghent Society members for the reproduction of parts of the study which had won the prize competition on the most useful hygienic measures in schools. The society did not hesitate to agree, but also stressed the obligation to mention the source of the study.[29]

Changing Hierarchies

Despite their track record of success, such staged expert performances in medical societies declined in the 1860s and 1870s. The reason for their demise was the gradual disappearance of the system of reciprocal legitimization between state officials and medical academics. In short, the combining of the roles of academic scientist and public health expert became less delineated as both fields expanded and became more independent. As a corollary of these changes, the different audiences of public health experts – medical practitioners, state officials and occasionally the general public – equally became more difficult to reach simultaneously. This inability resulted in the gradual fading of the philanthropic expertise of the middle of the century.

To contextualize this evolution, the newly professional group of public health professionals or 'hygienist physicians' (*médecins hygiénistes*), as they presented themselves, needs to be scrutinized. Among them could be counted the rising number of physicians working in government service, for example as 'doctors of the poor', providing medical service in working class districts, but also as physicians in prisons, school doctors and employees of the emerging urban public health services. These functions used to be taken up part-time by private practitioners, but gradually became full-time positions in the second half of the century. Government investments on the local and the national level, in food safety, sanitation and medical inspection were therefore crucial for the professional development of this medical group. These investments, and the new state services they created, also made state officials less inclined to outsource questions on public health to private medical societies. The basis for sustained collaboration thus gradually vanished. In addition, these hygienist physicians were also highly active in setting up larger professional organizations, such as the Fédération Médicale Belge in 1863, which would defend the interests of all physicians, and professional journals such as *Le Scalpel* which aimed to reach a wide medical audience with professional news. The organization of the medical profession thus also became more based on the defending of interests than on science and philanthropy.[30]

As a result of these developments, university professors partly lost their position as the unchallenged leaders of the medical community. They in turn focused more explicitly on scientific research. At the same time as professional organizations were being founded, the research university gradually developed and calls for 'pure research' were voiced by leading medical academics such as the Brussels professor Jean-Hubert Thiry.[31] The scientific community thus became more exclusive and academic. This did not mean, however, that university professors made no more attempts to reach a wide audience and claim a social relevance for their scientific work. On the contrary, together with scientific exclusivity, the popularization of science increased.[32] The emergence of a popular market for hygiene studies also meant that an audience of state officials was no longer the only audience which could be addressed to display social engagement. The interests of a burgeoning middle class also turned hygiene studies into commodities, of which the popularity reflected on the status of the author as a public health expert. This evolution is best illustrated in the careers of professors such as Burggraeve. From the middle of the 1850s, and increasingly in the 1860s and 1870s, Burggraeve published hygiene treatises and manuals. Many of these had multiple editions, his most successful one being *L'art de prolonger la vie.*[33] Individually published textbooks from the 1860s onwards turned out to be better vehicles to take up philanthropy and social engagement. These individual efforts, together with the heightened focus on scientific activities in medical societies, contributed to the view that these societies were not the places where such popularization should be practiced. For academics, the growing market of hygiene education and general treatises on health became a more suitable arena to take on a social role. The result was that the discourse of philanthropy largely disappeared from medical societies.

In the second half of the century, the medical community thus became more organized alongside the borders of scientific, professional and popular efforts. From the side of the academic community, this evolution should be viewed as a repositioning in terms of scientific exclusivity within a medical world where the borders between these fields were being negotiated. Within this changed context, the theatrical expert performances of the middle of the century appeared as a somewhat outdated form of social engagement. Addressing a public of both medical colleagues and state officials within the ever more exclusively scientific setting of urban medical societies seemed no longer possible. Such audiences required new strategies. This evolution, however, did not mean that the part of university professors in the development of public health expertise was played out. The changing hierarchies within the medical community also enabled new roles for medical academics and new expert performances in medical societies.

Review and Quality Control

The atmosphere of scientific rigour and exclusivity in urban medical societies was fundamental for the development of these new roles. While such exclusivity was most of all intended to encourage the scientific research at medical faculties and augment its prestige, university professors were not the only group who benefited from presenting themselves as 'scientists'. Because of their restrictive approach, medical societies also offered opportunities for the emerging group of public health professionals. These hygienist physicians similarly faced the problem of an expanding and diversifying market of works on hygiene. How to attract attention? How to legitimize one's expertise? To this end, both the membership of scientific medical societies and the reviewing of their studies by university professors were much sought after by hygienist physicians. Academics themselves thus evolved into a specific, scientific audience.

Such an audience was nevertheless difficult to reach for hygienist physicians. Scientific exclusivity offered opportunities to distinguish oneself, but equally made it difficult for hygienist physicians to enter, as their studies also aimed at an audience of policymakers. In the Medical Society of Ghent, the increase in exclusivity and focus on scientific medicine was outspoken. The society, even more than its counterpart in Brussels, became strongly connected to the university from the 1860s onwards. This academic turn was reflected in more restrictive approaches to membership and medical scholarship. In practice, mostly aspiring academics joined the society in the 1860s and 1870s. The functions of *préparateur* of the courses of anatomy or histology at the university or assistant-surgeon to the professors in the urban hospitals, for example, became stepping stones for an academic career. Physicians appointed to these posts were eager to join the city's medical society.[34] Hygienist physicians, to the contrary, rarely succeeded in becoming successful members. The intense reviewing of a growing number of medical studies held the inclusion of public health professionals back. Their studies not always met the scientific standards applied by the academics of the Medical Society of Ghent. Medical societies also functioned as the gatekeepers of the scientific community.[35]

The trajectory of César-Alexandre Frédericq illustrates these difficulties. In 1862, Frédericq became a member of the Medical Society of Ghent after a positive review of his work on the treatment of ophthalmia, a disease which had reached epidemic proportions especially among soldiers. Yet in the following year, while he was further developing his position in the field of public health, his work on hygiene in the hospitals was heavily criticized. This was not because such general studies of public health were deemed unfit for discussion in the society. One of the reviewers argued just the opposite, stating that because of the importance of the question, 'Any work on hygiene in the hospitals will be welcomed by us favourably'. But he also added: 'we make only one reservation, which is that we also recall that the desire to be charitable might lead to exaggeration'.[36] Further in the discussion, however, it was not so much with exaggeration as with vagueness and lack of originality that the reviewers blamed Frédericq's study. References to the authority of the most respected institution in medicine, the Parisian Académie de Médecine, were included to support such a claim. Had the necessity of separate pavilions in hospitals, as suggested by Frédericq, not for a long time been shown by the professors in Paris? Moreover, his work was not a concrete plan, but rather a collection of 'very general' ideas.[37] Such reviews functioned as mechanisms to exclude popular and philanthropic studies from the scientific efforts in the society.

After this review, Frédericq never again submitted his work to the Medical Society of Ghent. Yet, his somewhat failed scientific participation in the society did not mean his career in the field of public health was over. In the following years, he was successful in publishing different textbooks on matters of hygiene intended for a broad audience.[38] In the afterword to his *Handboek van gezondheidsleer* (Textbook on Hygiene), which was awarded a prize by the provincial council of East Flanders, Frédericq reflected on these different audiences, warning his audience of lay readers that while textbooks on hygiene could be read by anyone, studies of medical treatments should be limited to the medical community.[39] Frédericq himself certainly had understood this division in genres, focusing clearly on a general audience. His appearance at the Medical Society of Ghent was but a small intermezzo in a successful career of popularization of knowledge on hygiene.

Frédericq's example also shows that for hygienist physicians hoping to convince an audience of university professors, different strategies were needed. The innovative value and scientific character of public health studies rather, was to be stressed. Expertise therefore became performed in more subtle ways in medical societies. Instead of 'grand' performances that highlighted philanthropy and social engagement, reviewing practices became means for young, aspiring hygienist physicians to show their scientific competence. The discussion by one of these young hygienists, Émile Vandermeersch, of studies on the treatment and prevention of rabies, reveals how reviewing was also a way of establishing authority. After becoming a member of the society in 1868 – the result of a positive review by professor Charles Van Bambeke of his short article on an observation of rabies – Vandermeersch reviewed a study by Raucq, a physician from the small town of Loochristy. In his evaluation, Vandermeersch showed himself highly critical of Raucq's work, '[which] was nothing more than the incomplete reproduction, and in certain passages, in servile manner of a small brochure ... by Jac. Dycer'.[40] By unmasking Raucq's study, Vandermeersch in turn showed his mastery of the rapidly growing body of literature. Similar judgments were made in the field of cholera studies. Here, the challenge of mastering the literature was even more present. Some works were immediately dismissed, so meeting reports show, since they were said to contain nothing new.[41] These practices reflect the setting of boundaries between scientific and popular expertise in public health and the changing values, such as mastery over the literature and innovation, which were attributed to public health experts. By displaying such values, aspiring experts such as Vandermeersch tried to distinguish themselves from authors such as Raucq and Frédericq. They presented themselves as scientists instead of philanthropists since performances of charitable outcomes were more and more regarded as unscientific.

The medical society – with its scientific audience – provided indeed the right forum for such an undertaking. Even if these societies were no longer major players, they nevertheless played a part in the development of public health expertise. Within a changing medical landscape, the more exclusive interpretation of 'science' in these societies formed a means of quality control for such a new field in which the boundaries between popular, professional and scientific work were in the making. Such roles of arbitrage therefore seem to confirm what Stéphane Van Damme has called the growing need for 'expert judgment' in an expanding market for knowledge. The 'production of norms', Van Damme argues, was fundamental for the development of expertise.[42]

Prestige and Reputation

In Brussels, the admission of hygienist physicians in the city's medical society followed a different trajectory than in Ghent. The members of the Brussels Society were more receptive to the studies conducted by this new professional group. Several reasons for this difference can be distinguished. First, the foundation of the Société Anatomo-pathologique in 1858 provided the professors of the Brussels university with their own specific forum for presenting clinical cases, leaving more room for general medical discussions in the existing Société des Sciences Médicales et Naturelles de Bruxelles. Second, the presence of institutions of the central government, such as the Académie de Médecine, made Brussels more into a centre for public health expertise than Ghent. The worlds of science and politics were thus more closely intertwined, making public health into a crucial theme for the Brussels Society.

The interest in public health research was first of all reflected in the less exclusive distinction that was made in Brussels, compared to Ghent, between scientific and non-scientific studies. This did not mean that the Brussels professors did not distinguish between scientific studies and what they called 'administrative' studies, which discussed the organization of health services by the state. It rather meant that this division did not always determine the acceptance or refusal of studies. When the pharmacist and Doctor of Sciences Théodore Belval submitted his study on the organization of public health in Belgium in 1872, professor Louis Martin argued:

> This study ... belongs more to the administrative domain than to the field of proper science. Yet, public health, the object of the study, has an importance too marked for our Society not to be interested in all measures that help spread its application and certainly its organization.[43]

Martin's judgment was typical of the way studies of public health were treated in the Brussels Society. Reviewers always had to compromise between the scientific value of the study and its social importance. So even if studies were thus openly characterized as non-scientific, they could still be discussed. Belval, as a result, became a member of the society.

For hygienist physicians, emphasis on the social importance of their studies of public health formed a way of convincing an audience of university professors. Eugène Janssens, physician, statistician and head of the Brussels health services, was particularly skilful in placing matters of public health on the agenda of the society.[44] As his fame rose on the international scene, especially after the international Hygiene Conference of 1876 in Brussels, he became increasingly successful in tying his own studies to the pioneering role that the Brussels Society wanted to play as the 'leading' scientific institution of Belgian medicine. Janssens, for example, compared the health services of Brussels and Paris in 1882, and argued that Brussels was enjoying 'an epidemic of good health' because of the work of the city's health services.[45] Such statements served his own ambition of spreading the Brussels model of organization, but equally stressed the pioneering role of the Brussels medical community. International comparisons were indeed particularly fit for such diverse goals. In 1878, Belval had made a similar remark concerning a report by the French physician Octave Du Mesnil on the conference of 1876. In his review, Belval highlighted that Du Mesnil viewed the organization of public health in Brussels as a model for Paris.[46] With such statements, academics, hygienist physicians and even an audience of state officials could be reached.

The admission of hygienist physicians, however, did not always pass off smoothly. Different interests were sometimes hard to reconcile. The affiliation of Hubert Boëns to the Brussels society forms an example of such a difficult integration. In 1860, Boëns had sent his study on the organization of public health in the region of Charleroi to the Brussels Society. The same reasoning as regarding Belval's study was followed. The reviewer argued that the work 'had no scientific value and comprized exclusively of recommendations and views on the organization of medical services'.[47] Yet, Boëns nevertheless became affiliated to the society as a corresponding member. The members of the Brussels Society, however, would later regret their decision. Boëns, unlike Belval, was a rather controversial figure, who took on sharp positions in contemporary debates, including the debate on vaccination in the 1860s, of which he became one of the most well-known opponents.[48] The study, reviewed in the Brussels Society in 1860, formed no exception to his trajectory of controversy. A group of Charleroi physicians reacted against the review in an open letter, claiming that 'the elevated and scientific atmosphere' of the society had failed to pick up the flaws and personal interests behind Boëns's so-called objective description.[49] The Brussels professor Crocq tried to put the matter into perspective by pointing to Boëns's reputation for 'writing too much and thinking too little', but the letter of protest had painfully made clear how the reputation of the society could be damaged by unintentionally engaging in professional debates through the allocation of memberships and the review of public health studies.[50]

Such controversies raised the consciousness of the society members of their role in legitimizing expert studies as scientific works. By judging studies, and their authors, the society members not only interfered in the development of public health, but they equally engaged in professional disputes by reviewing some studies and dismissing others. The awareness of their arbitrational role lead to debates among society members on their responsibilities as editors of the society's journal. Were they responsible for the opinions articulated in the articles published? In a heavily discussed review of a cholera study by Loneux, a rural physician, the chairman ended the discussion by stating:

> To sum up, I think the work by M. Loneux can be published in our journal, because the discussion ... [as will be printed in the meeting report] will make sufficiently clear to our readers that we do not accept the opinions of the honourable physician ... without reservations.[51]

Such discussions were common practice. They show that the role of medical societies as 'arbiters of science' also exposed these societies and their members to criticism of their authority.

Expertise and Politics

Despite such problems of arbitration, several hygienist physicians had become members of the medical societies of Ghent and Brussels by the late 1870s. By participating in the scientific community, they had gradually replaced a philanthropic approach to public health with a scientific one. The question which came up in the 1870s, as a corollary of this transition, was whether an audience of politicians and administrators could still be explicitly addressed. To what degree were public, engaged expert performances, such as Burggraeve's speech in 1864, still feasible when they were staged by public health professionals stemming from an exclusively scientific setting? Here again, societal relevance and scientific exclusivity had to be reconciled. Yet, these debates on public engagement, even more than reviews, also reveal the fragilities that emerged together with the 'scientization' of public health expertise.

In the Medical Society of Brussels, such questions of public engagement were greeted with scepticism. Its academic members tended to argue in general that the society's scientific ambitions excluded such public positioning. Such argumentation was, for example, advanced by Jean-Hubert Thiry on the issue of cremation. In the 1870s, the regulation of cremation had led to a heated debate between liberal and catholic parties, in which not only ideological opinions, but also arguments of public health were used. Within this polarized context, Thiry tried to prevent the society from taking on a more elaborate public role and argued that

> the matter would be no longer to limit our intervention to a purely scientific role, we would have to exert political influence, vulgarize and defend this new system [cremation] through the press, contact various governmental powers and provoke a general petitioning.[52]

Such tasks, Thiry suggested, were not essential to the functioning of the society, and should therefore be avoided. In the Medical Society of Ghent, several studies of the topic were also reviewed, but similarly to Brussels no attempt was made to intervene in the public debate. At the furthest, a potential audience of policymakers could be taken into account during a discussion. Professor Victor Deneffe, for example, criticized a new technique of inhumation in stone coffins because of its high financial costs, which made that the technique had 'little chance of being adopted by the administrators'.[53] Such statements, however, were rather scarce. In the late nineteenth century, medical societies seemed to focus on their core business: the practice of science.

At the basis of the reserve of medical academics, however, was more than merely scientific exclusivity. Their hesitant position should also be interpreted within the politicized social debates of the 1870s. In these years, political tensions emerged between progressives, advocates of a stronger intervening role for the government and conservatives, adherents of limited government intervention. Such tensions extended into the medical world, which equally became more politicized in the second half of the century.[54] The academic members of the Medical Society of Brussels mostly belonged to the conservative group. During the nineteenth century, the city's university functioned as a breeding ground for liberal ideology. But while progressive views became popular among medical students, they only gradually seeped through at the medical faculty, where conservative views continued to hold sway.[55] In the 1870s, several members of the medical faculty took on political mandates. Jean Crocq was elected a member of the Belgian senate in 1870 and a member of the provincial council in 1872. Arsène Pigeolet was a member of the Belgian senate and the town council of

Brussels.[56] Their views of individual liberty and limited state intervention were shared by their academic colleagues. Such consensus among university professors might have inspired president Thiry to declare, in his opening speech in 1870, that liberalism was at the core of the Belgian nation and therefore should also form the basis of Belgian medicine.[57]

Within such a political context, the pleas for more government intervention by mostly progressive hygienist physicians could only attract limited support. Hygienist physicians and medical academics disagreed on various topics of state intervention. Public mortuaries, obligatory vaccination and the promotion of breastfeeding were contested issues. For the Brussels academics, mortuaries were without doubt hygienically better than placing bodies on the bier at home, but not strictly necessary as the local doctor could also make the necessary arrangements and determine the cause of death. Breastfeeding was unmistakably beneficial for each child, but rendering it mandatory a bridge too far. 'Liberty, nothing but liberty, these are our principles', the meeting report documents, 'touching upon the liberty of the family that is a question which will not gather many partisans in this circle'.[58] As a corollary of these conservative liberal views, no support was given to the proposals of hygienist physicians to collectively advocate health reforms. Janssens's plea for the construction of mortuaries, for example, evoked little interest.[59]

The differences between both groups were profound. At the basis of their political disagreements were also different views on the relation between the physician and the state. For the Brussels professors, the independent expert was the logical counterpart of the limited, liberal state. The government, in their opinion, should not establish extensive health services, but rather outsource questions to the private medical world. These academics opposed any systematic organization of salaried employment by the state, which would turn the physician into 'one of the radars of the great machine [of the state]'.[60] In debates on the organization of public health, they pleaded instead for more financial support for private associations and defended the functioning of the Académie de Médecine. Here again the medical world was divided between hygienist physicians, mostly progressives for whom a salary by the state was essential, and academics who defended the view of medicine as a 'liberal profession', free from state control.[61]

Yet the power of the Brussels professors should equally not be overestimated. Their rigid views on the independence of the medical profession also seemed somewhat outdated in the late nineteenth century. This was reflected in the isolated position of their society within the medical landscape. Emphasis on autonomy and independence had led to refusing participation in over-arching networks. In 1874, and again in 1876, a request to join a federation of scientific institutions was rejected. Some members were worried that the established name of the society would be damaged by participating in a network of institutions of which the scientific nature was not always clear.[62] After a dispute concerning a law proposal, the society also broke all ties with the Fédération Médical Belge, even though Jean Crocq had been president of the organization for ten years, between 1863 and 1873.[63] While these professors indeed stood relatively isolated, their reserve nevertheless hints at some of the fragilities of combining scientific research, expert advice and political statements. For the Brussels professors, such combinations jeopardized the autonomy and credibility of the medical expert. Unlike the advice of academics who were neither full time experts, nor paid officially, the expert performances of hygienist physicians seemed to entail a conflict of interest between experts and the state.[64] Hygienists' credibility could easily be damaged by pointing to the self-serving motives of their pleas for more government intervention, which would also create more professional positions for hygienist physicians. For the Brussels professors, their reputation as scientists outweighed the risk of being accused of such motives by engaging in public debates.

These political and professional differences certainly made public statements into a difficult matter for medical societies. Yet there were also possibilities of compromise. Agreements could always be found on the 'scientific' side of the question. In the debate on vaccination, for example, the scientific evidence of the benefits of vaccination was clear to all members, unlike the need to render vaccination mandatory. This agreement, in turn, led to a public statement of the medical society. At the end of the meeting all members collectively proclaimed their confidence in the practice of vaccination.

From Agreement to Engagement

Such statements were forerunners of new expert performances in the late nineteenth century. Scientific consensus could remove fears of interference in public debate, safeguard reputations and justify public actions. The 'scientization' of the public health debate was thus also a strategy of compromise. It enabled a collective, public engagement by medical societies as authorities of science.[65] Put differently, scientific arbitration could also be extended into the public debate.

The negation of rumours concerning epidemics forms a clear example of such public scientific statements. During a discussion on reigning diseases in 1867, the Ghent medical society declared that 'the sanitary condition of the city was satisfactory and that nothing justified the alarming rumours caused by superficiality or fear'.[66] In Brussels in 1882, President Joseph Sacré asked his colleagues during a similar discussion whether they had any knowledge of incidences of cholera. During the last meeting of the town council, such incidences had been denied by the mayor in response to rumours of cholera in the city. Some members had viewed cases of *cholérine* (of diarrhoea), but cholera proper, so they declared in group, was not present in the city.[67] Such judgments made medical societies into scientific authorities because of their shared, collective knowledge.

They could also unite the different groups within medical societies. In a discussion on vaccination, for example, Professor Thiry recast the distinction between matters of practical and scientific importance in a new, positive way, arguing that 'our society, where one finds authoritative men, both from a scientific and a practical perspective, is in a perfect position to clarify the question'.[68] It was precisely the diversity of the society members, including in particular the public health professionals who had succeeded in becoming members – those 'practical' men – that made it possible to exert scientific authority in a public debate.

Another form of displaying scientific authority in public debate was the publication of sanitary bulletins. These were equally the result of compromises. The matter of cremation, for example, was settled by creating a new section in the society's journal. Under the influence of some of the public health experts, it was decided that the Brussels Society could report on new techniques and scientific studies regarding cremation during its meetings. This information would then be published in the new section.[69] Other sections were also added to the journal. Statistical tables on the causes of death in the city, sanitary bulletins during epidemics,[70] and general hygienic and climatological tables were gradually added or expanded by hygienist physicians. It was, for example, Janssens who created mortality tables, on the basis of his work in the city's health services, and published them in the society's journal. These tables and overviews should be regarded as the result of negotiations, of the widening of the function of being a 'scientific authority' in the public debate.

Finally, the forwarding of medical studies to the urban government can similarly be considered as a public expert performance. In the Medical Society of Ghent, such actions certainly became common in the late 1870s.[71] Here aspiring hygienist physicians, such as Eugène De Keghel, collaborated with professors in chemistry, active in the emerging field of bacteriology, such as the physician and Doctor of Sciences Nicolas Du Moulin. In 1877, De Keghel had criticized the use of lead in the installations of breweries, which could lead to intoxications. After analyses by Du Moulin, and with the approval of his colleagues, he drew the attention of the urban government to the matter. According to De Keghel, the most basic scientific notion needed to be emphasized. Lead, in any construction, always remained toxic. Such information formed 'the best advice for anyone interested in public health'.[72] Here again, an audience of policymakers was reached. Such actions can be understood as reinventions of older expert performances, which were now more rooted in science and actively addressed to an audience of state officials, rather than presented as forms of philanthropy and delivered on demand.

Conclusion

In the second half of the nineteenth century, 'philanthropic' elite practitioners had gradually been replaced by scientific professionals as the experts in the field of public health. This chapter has analyzed this transition by looking at the 'performances' of these experts, and has examined, in other words, the strategies used by them to convince different audiences of their authority, and to acquire an expert status. Such a methodological approach offers a new understanding of the professionalization of public health, not so much as a process of collaboration between two actors – physicians and the state – but rather as a multifaceted process, in which old and new experts interacted with each other, medical interests and scientific research had to be balanced, and a constant repositioning of the physician within the expanding health policies of the state was required. The connecting thread within these complex changes was experts' search for the right audiences and the right means to reach them. As the relatively small-scale, civic medical world of the mid-nineteenth-century experts, in which a philanthropic framing of expert studies satisfied state officials and brought social prestige among colleagues, was replaced by an expanding and diversifying medical field, new forms of legitimization of expert authority became required. Within this changing landscape, science was used as a marker, as a means to distinguish valid from non-valid expert knowledge.

From a general perspective, one could argue that while the 'framing' of expertise in public health changed, its basis – the advisory role of the physicians vis-à-vis the state – remained constant throughout the century. Such continuity certainly needs to be recognized, but, at the same time, it is also important to point to structural changes in the relation between experts and the medical profession, and between experts and the state. The mid-nineteenth-century expert, in fact, stood at the top of the profession. More than that, his status as a leading member of the profession made him precisely into a trustworthy expert in the eyes of the government. The taking up of advisory roles was in many ways thus a complement to an elite status. Both reinforced each other. The late nineteenth-century public health professional, to the contrary, was not necessarily a member of the medical elite (although his work as an expert could form a road to professional success). His career often started out modestly, and only boomed as he managed to navigate in the expanding state infrastructure of public health. Within the medical community, he increasingly presented himself as a scientific specialist. This evolution continued into the final decades of the century. By the turn of the twentieth century, 'hygienism' would develop into a scientific specialism among the range of subfields that marked modern medicine.

As for the relation between experts and the state, the changing 'framing' of expertise is indicative of deeper lying changes in the social and political views of experts. From this perspective, philanthropy was an appreciated attitude for the gentleman scientist, who through his expert work displayed civic engagement, but at the same time also patronized the emerging social movement of the middle of the century. Science on demand, in which expertise was an honorary undertaking for engaged citizens, was the logical complement to a limited state that left the solving of social problems to a large extent to civil society. Within such a constellation, experts were not at all part of state machinery. By the late nineteenth century, the presentation of public health studies as scientific works reveals a different socio-political logic. It was based in scientism, in the sense that the proposed, scientifically grounded measures were to transform society, rather than alleviate suffering. Public health professionals typically adhered to the more progressive factions within political (often liberal) parties, and became much more immersed in politics and state machinery than their predecessors. The distance between experts and their employers thus narrowed in the second half of the century, and with this narrowing came a problem of independence and reliability, which led to discussions in the medical community and could be overcome, to a certain extent, by collective authoritative statements.[73]

3 BORDERLESS NATURE: EXPERTS AND THE INTERNATIONALIZATION OF NATURE PROTECTION, 1890–1940

Raf de Bont

In 1999 the Pan-European Ecological Network, which operates under the auspices of the Council of Europe, held an international symposium with the title 'Nature Does Not Have Any Borders'.[1] This title, in fact, was far from original. By 1999 it had become a truism in conservation circles to proclaim that nature did not take national borders into account. If something was truly international, if something incarnated the common heritage of mankind, it was considered to be nature. It is a reasoning to which most of us have become accustomed.

If in our present-day culture nature is given a residence, it is often the globe as a whole. Particularly with regard to threatened nature, the image of the earth has received an iconic status. The 'blue marble' – the picture the crew of the Apollo 17 took from the earth in 1972 – was instrumental in this. It shows the earth as a fertile, but also as a vulnerable place in a dark universe. Environmental groups all over the world have borrowed the image. They have also drawn on the idea spread by scientific ecologists that organisms across the globe are closely interconnected and interdependent – an idea that probably took its best-known and most controversial form in James Lovelock's Gaia theory. It is the same idea of interdependence that has been used by activists defending the cause of the world's biodiversity, or that has made them portray the Amazon forest as the lungs of the earth. The recent public discussions about global climate change have, of course, only strengthened the discourse of the internationality of nature.[2]

Yet the idea that nature does not have any borders is not self-evident. It has been historically constructed. Around 1900 the associative field in which nature was situated was mostly of a local and a national kind, and early nature protectors were generally eager to stress the patriotic value of their work. It was only alongside these patriotic protectors that the early twentieth century witnessed the rise of an internationally oriented conservation movement.[3] It is in this period that the disappearance of natural areas and the extinction of species was increasingly defined as an 'international' problem. In the same years, the problem was also ever more portrayed as a 'scientific' one. A relatively small network of experts was instrumental in this shift. Their expert role was co-constructed with the image of nature's internationality, roughly in the years between 1890 and 1940. Despite its lasting influence on the way we think about nature, this development remains underexplored by historians.[4]

This article studies the double construction of the objects *and* experts of international nature protection. As such, attention will be paid to the institutional, epistemic and social constellations that brought forward both 'international nature' and the specialists of its conservation. In the period under consideration, it was naturalists (particularly zoologists), who were most successful in taking up the expert role. But they were far from the only candidates. Also foresters, hunters, colonial administrators, jurists and even travel writers and open air painters could claim knowledge on the matter – and they effectively did. If zoologists were to take up a leading role in international nature protection they had to engage in boundary work. They had to convince others of their expert role, and clearly delineate the boundaries of their expertise. They had to define which nature was worthy of protection, offer knowledge with which this protection could be achieved, and persuade an audience (of both fellow-scientists and policymakers) to take over their recommendations. This essay studies how these varied practices shaped the early years of international nature conservation.

Geographically, this article focuses on the European engagement with international nature conservation. Evidently, also the United States played a significant role in this period – a role, for that matter, which has been documented by historians in quite some detail.[5] In the interwar years, the Americans were involved in several international projects of nature conservation as initiators, lobbyists or benefactors. At the same time, transatlantic contacts were less intensive than those within the respective American and European networks. It is this last network that takes centre stage here.

Rooted in Local Ground

In the period around 1900 international projects of nature protection were still modest, and largely outshone by initiatives taken on the national level. It is in the same years, after all, that in several European countries the nature protection movement underwent a spectacular growth and a strong institutionalization. In 1894, the National Trust for Places of Historic Interest or National Beauty was founded in the United Kingdom; in 1901 the Société pour la Protection des Paysages de France; in 1904, Germany's Bund Heimatschutz. In these organizations it was not scientists, but rather literary intellectuals, poets and social reformers who initially took the lead. Mostly, their focus was on the protection of aesthetically pleasing and historically engaging nature.[6] Yet, rather quickly, naturalists would also take the scene and complement the artistic agenda with a scientific one. Probably the most successful in this was the German botanist Hugo Conwentz. His work can also serve as a good example of the interweaving of nature protection, national sentiment and naturalist expertise at the beginning of the twentieth century – and thus be helpful in painting the cultural backdrop against which international conservation initiatives took off.[7]

Conwentz's influence derived from his position as director of the Central Institute for the Care of Natural Monuments, which was founded in 1906. The concept of 'natural monument' was crucial to the vision of nature protection that Conwentz developed for his institute. It was a sufficiently vague term that could include beautiful sceneries and views, noteworthy rocks and characteristic soil formations, single old trees as well as vegetation groups and rare animals.[8] As such it could accommodate the artistically inclined nature protectors as well as the naturalists. Furthermore it was in line with the successful late nineteenth-century tradition of *Heimatkunde* (homeland studies) – a branch of learning that merged popular nature study with local history, folklore and conservation. The term 'natural monument' (a metaphor that provided nature with a counterpart for man-made historical buildings) also explicitly tied the natural world together with the human past.[9]

But Conwentz also brought a specific angle that came from the natural sciences. In his inclusive approach to conservation one discipline served as the backbone: plant geography. The geographical distribution of plants constituted a particular point of interest for German field botanists ever since Alexander von Humboldt had put the discipline on the map at the beginning of the nineteenth century.[10] Around 1900, plant geography witnessed a revival under the name of phytosociology, which revolved around so-called plant associations: groups of associated plant species that can be found in correlation with certain soil types and climate zones. By inventorying plant associations one could delineate the natural regions of a country, which served an important role in Conwentz's plans of protecting the diversity of Prussia's nature. Plants were literally rooted in local ground and thus became the true markers of the *Heimat* (homeland). Inventorying them, and in this way tracing the sites most suited to put up reserves, was one of the main aims of Conwentz's institute. The fact that plants do not (or hardly) move around was obviously helpful. Free-roaming animals lent themselves less to the approach.[11]

Conwentz's vision was based on pinning down nature geographically, and, thus, making natural monuments the unmovable incarnation of the national and the local soul. In 1904, he stated: 'Through natural monuments we will protect and save meaningful lands of our confined *Heimat* as well as our German fatherland, and as such, these efforts have besides their scientific and general meaning also a strong *national* importance'.[12] Such ideas were not limited to Conwentz and his circle. In the German Reich several plant geographers took part in the national nature protection movement, and also in Belgium, the Netherlands, France, Switzerland and Denmark they propagated plant associations as the organizing principle for a scientific type of nature protection.[13] Their work was not without success, and even outside the work of botany, plants increasingly became markers of identity. In 1916 the Dutch folklorist Josef Schrijnen, wrote: 'As long as one can still find traces of the natural vegetation in the Dutch forests ... we will not despair about the conservation of our nationality and the old Dutch national character'.[14] It was a direct echo of Conwentz's ideas.

The nationally oriented movement initiated by Conwentz was associated with a particular institutional context. It became a success thanks to Conwentz's national institute and the local societies that were related to it. Yet, around the turn of the century, nature protection would also become a topic of discussion in a completely different setting: the international scientific conference. It is in this context that nature would be defined as an international object.

Conference-Goers and Migrating Birds

In 1900 international scientific congresses were a relatively new phenomenon. Only in the second half of the nineteenth century did they grow into an established practice within the world of science. International conferences dedicated to naturalist issues sprung up even later. In 1884 the first international conference for ornithology was held (in Vienna), in 1889 the first for zoology (in Paris) and in 1900 the first for botany (again in Paris). These gatherings were structured around a series of lectures, discussions, banquets and excursions. They were used to agree upon scientific standards and to shape disciplines; and they offered occasions for networking both with fellow scientists and with the people of power (who were usually represented in the patronage committees). Furthermore, the conferences grew into powerful symbols of the internationality of science. As such, they constituted a fertile ground for internationalizing the issue of nature protection.[15]

It was at the ornithological conferences that the conservationist concern first crystallized. The ornithologists who gathered there were alarmed by declining bird numbers, which they attributed to worldwide hunting and a growing feather industry specialized in adorning fashionable women's hats. The international stage was considered the best place to discuss the matter, not only because the plume trade was a world-spanning business, but also because the objects of concern could be seen as 'international travellers'. Migratory birds were the opposite of plants – they could not be pinned down to a certain *Heimat* as they easily crossed the borders between nations and continents. The ornithologists called for their protection and they presented themselves as the key experts to make this protection work. Two types of knowledge were needed to make bird preservation scientific, so they claimed. First one needed to chart the bird's migratory routes – a project similar to that of the plant geographers. The international conferences were used to coordinate this charting, first by setting up an international network of observatories, later by organizing large-scale projects of bird ringing.[16]

Yet, according to the ornithologists, scientific protection not only involved tracking birds, but also assessing their relative value. The ornithologists, after all, did not set out to protect *all* birds, but only those that were considered to be useful – and it is with regard to this point that they could advertise their knowledge as being of societal importance. Late nineteenth-century Europe witnessed a series of insect plagues that hugely damaged agriculture and forestry. By analyzing which birds ate insects, the ornithologists hoped to enable an economically beneficial project of protection. One of the leading men behind this idea was the zoologist Hans Rörig, who worked for the German Biological Department for Agriculture and Forestry. Meticulously, he performed stomach analyses, looking for agricultural seeds (considered a bad thing) and insects (considered a good thing). By doing so, he presented himself as the bookkeeper of nature's economy. This bookkeeping project was copied in several other countries.[17]

Rörig's utilitarian accounting was not without its problems, however. These problems were highlighted at the conferences as well. At the international ornithological conference of 1900, one participant pointed out that birds often changed their diet when travelling to other places. Another indicated that it mattered which kind of insects birds ate, since some insects preyed on other insects and thus had to be categorized as useful rather than harmful. Still others argued that categories of usefulness were superficial anyway, since every species had its role to play in nature's equilibrium – a standpoint that would become increasingly popular later in the twentieth century.[18] But although these discussions concerned the fundament of the ornithologists' expertise, they seem not to have hampered their effectiveness. Shortly after the 1900 conference, and on instigation of the experts present there, a Convention for the Protection of Birds Useful to Agriculture was signed by twelve European countries.[19]

The ornithologists, furthermore, spread the issue of conservation to wider scientific circles, amongst others, by setting up sessions devoted to the topic at zoological conferences. Also in this context they portrayed themselves as rational and scientific protection experts. Otto Kleinschmidt, for instance, proclaimed at the zoological conference of 1904: 'Down with sentimentality! The preservation of the animal world is in our civilization only possible when it is carried out on the basis of scientific principles, which means that it is based on the work of professional zoologists'.[20] The boundary work goes in two directions. It served to broaden the group of scientific protectionists from ornithologists to *all* zoologists, but also to exclude everyone who was no professional. The latter was particularly meant to dissociate the protection cause from the British women's societies that in the 1880s had started to campaign against the feather industry, mostly by propagating more moral forms of fashion. The most active society, the Royal Society for the Protection of Birds, carried as one of its rules that 'Lady-Members shall refrain from wearing the feathers of any bird not killed for purposes of food, the ostrich only excepted'.[21] The ornithologists – or at least some of them – made sure to distance themselves from those supposedly hysterical activists, by claiming an expertise that was rational, masculine, and rooted in professional training.[22]

Paul Sarasin, Zoologist-Protector

The attempts at defining nature conservation as a matter of professional, zoological expertise came at a time period in which it was still unclear who was to act as the spokesperson of the world's threatened nature. The women feather activists and the male scientific zoologists were operating in a much larger field that was professionally very heterogeneous. A British conference in London in 1900 focusing on the preservation of African fauna, for example, was dominated by diplomats, colonial administrators and big game hunters. Independently from the ornithologists, who had their conference only a month later, these men drew up a document that breathed the same utilitarian vision of nature that inspired Rörig's stomach analyses. It resulted in a convention that hoped to protect wild animals 'either useful to man or ... harmless'. The only zoologist present at the conference mostly used his influence to avoid a call for the absolute extermination of so-called 'vermin' (a category which included species such as lions and leopards).[23] In 1909, then, it was a group of literary intellectuals, artists, lawyers and politicians who dominated the International Congress for the Protection of Landscapes in Paris. The one ornithologist on the speakers' list talked on the protection of wildlife, but his ideas got lost in an audience that was more interested in nature's aesthetics and *Heimat*-related questions.[24]

Within the hybrid group of actors involved in turn of the century discussions, the zoologists would start to sharpen their profile. By the early 1910s, they increasingly used the international zoological conferences as a stage for propagating a scientifically inspired, worldwide nature protection. Particularly active was the Swiss zoologist, ethnographer, traveller and gentleman scientist Paul Sarasin. At the zoological conference of Graz in 1910 he put the theme of 'world nature protection' on the agenda.[25] Three years later he set up an entire conference devoted to the theme in Bern, where a Consultative Commission for International Nature Protection was founded. The First World War and its aftermath led to a collapse of the Commission before it could take off, but Sarasin's ideas continued to be influential for several decades – arguably even until today. Amongst these ideas was his well-known statement that 'nature does not know political borders'.[26]

In the first address Sarasin devoted to world nature protection (in 1910), he firmly situated it in the realm of science. He indicated that he did not address the gathered zoologists as a 'stranger', but rather as a man who had devoted most of his life to scientific study. Through his work, he indicated, he had become convinced that nature protection counted 'also for the rigorously trained researcher as a primary duty'. Asserting his identity as a field naturalist he stressed that the time had come to take a look outside of the laboratory to behold the 'sad impoverishment of our beloved nature'. His aim was to counter this impoverishment through fostering a well-selected network of militants. The fact that he launched this idea for a strictly scientific audience already indicates that he was thinking of an expert-driven pressure group.[27]

His goal was one of identity-building: Sarasin had to convince the zoologists *themselves* that they were able to play an expert role in matters of nature protection. To achieve this aim a homogeneous scientific conference such as the one in Graz proved to be more fitting than the previous, more heterogeneous conferences in London or Paris.

Sarasin approached nature protection from a particular angle. Unlike his fellow-naturalist Conwentz, he did not focus on plants, but on animals. And he concentrated on those animals that were nationally ambiguous. Next to migratory birds, Sarasin devoted a lot of attention to the declining numbers of whales in international waters. He spoke of the Arctic and Antarctic no-man's land (and its threatened seals, polar bears and penguins); and he proposed to turn the disputed islands of Spitsbergen into an international nature reserve.[28] Furthermore, he paid a lot of attention to the tropical parts of the globe, particularly to Africa's great mammals. These mammals were known to migrate across the political borders the colonial powers had drawn, and, thus, could easily be presented as an object of international concern. The African fauna was furthermore believed to offer a more primitive, more unharmed aspect than the European one, and its protection was presented as a responsibility of the civilized world as a whole, rather than that of individual colonial states. This responsibility, finally, did not only include animals, but also so-called *Naturvölker* ('primitive' indigenous populations). As parts of primitive nature, they were to be protected against destruction.[29]

When addressing the threats to primitive nature, Sarasin used a tone that was certainly more moralizing than had been common among zoologists. He broke with the utilitarian bookkeepers' attitude and spoke of protecting the harmony of nature. In order to preserve nature's harmony, he came up with a two-sided plan of action. He argued for the protection of single species (like for example the diplomats of the London Convention had argued for), but he also hoped to protect entire associations of organisms (thus echoing the protectionist agenda of Conwentz). Central in the latter strategy was the concept of 'life community' or 'biocoenosis' – a notion that referred to the whole of interacting organisms in a certain habitat.[30] Sarasin aimed for protecting these life communities by setting up so-called total reserves, which should be free from all human influence and preferably cross national borders. Apart from scientists, Europeans would have no access to these reserves, while the included primitive tribes would be forbidden to *go out*. As such, the reserves would constitute 'living laboratories', which the scientists could use for the study of primitive nature, or in Sarasin's words: *Urnatur.*[31]

Sarasin's conception of conservation radically cut off western man from nature. While artistically inspired landscape protectors or plant geographers focused on the *Heimat* and local history, world nature conservation in the hands of Sarasin was about protecting the Other. His interest concerned 'wild nature' (a harmonic whole of soil, flora, fauna and non-western humans) that could serve as the counter-image of the European metropolises in which the zoologists held their conferences.

Interwar Networks

Despite its international rhetoric, it is not very difficult to see that Sarasin's project had to be accommodated within *national* agendas. It was financially backed by influential Swiss politicians who hoped for an increased role of Switzerland in international post-war diplomacy. The German government, for its part, followed a policy line that above all stressed national sovereignty, which partially explains Conwentz's lack of enthusiasm for Sarasin's commission (next to his predisposition to protect the internationally leading status of his Prussian institute).[32] And even in Sarasin's own mind the international project sometimes involved that national borders should be strengthened rather than overcome. Repeatedly, he argued in favour of extending the territorial waters to come to a complete national division of the world seas. This, he believed, would make it easier to protect the worldwide whale population.[33] Yet, rather than furthering Sarasin's plans, national division lines would eventually undermine it. Following World War I national sensitivities were responsible for the failure to reestablish his consultative commission under the auspices of the League of Nations.[34]

After 1918, it was only through shuttle diplomacy that a small informal network of scientists kept the international spirit alive. In 1923 an international conference for the protection of nature was held in Paris – an event from which the Germans were excluded and which was dominated by the French.[35] In 1931, a second, more inclusive Parisian conference was held, to which a German delegation also attended.[36] Simultaneously, the Poles and the Belgians tried to promote the Union of the Biological Sciences as an international forum for discussions on nature protection, while the British similarly propped up two conferences for the international protection of the African fauna.[37] In the same years the first permanent (but very small-scale) institutions for international nature protection were set up. In 1922 the International Committee for Bird Protection was founded, followed in 1928 by the Office International de Documentation et de la Corrélation pour la Protection de la Nature. The two institutions shared a secretariat in a Brussels townhouse.[38]

Throughout these different institutional settings one could witness the rise of the scientific expert that was so typical for late modernity. The Paris conferences, for instance, were built on the pre-war Congress for the Protection of Landscapes, but had a rather different composition. The landscape painters and politicians were now clearly put in the shadow of the zoologists and the agricultural engineers. The organizers of the conference had been inclusive enough to present the nature protection movement as a broad alliance, but strategically made the rhetoric of science its dominant language.[39] Similar developments can be seen in the London conferences of the 1930s. These conferences took up the issue of African fauna where the 1900 convention had left it, but while in 1900 only one scientist had been present, now most diplomats came with an entire commission of scientific advisers. At the 1933 conference, the British delegation alone counted fourteen of them.[40]

The form of the interaction between scientific experts and policymakers differed from conference to conference. The Parisian conferences of 1923 and 1931 tried to influence governments by having the participants vote on resolutions (voeux) that asked for particular reforms. The London conferences of 1933 and 1938 saw a more integrated approach in which diplomats and experts closely interacted in preparing texts for an international convention. In both cases, however, it was by directly addressing people of power, rather than campaigning for a general audience, that conservation experts tried to influence policymaking. Informal contacts with diplomats, ministers and aristocrats were crucial in this. Next to the conferences, also balls, dinner parties and royal audiences offered occasions to foster and maintain these ties.[41]

The international ramification of the expert network offered a strategic benefit for convincing policymakers – also when the agendas of the latter were to a large degree national or local. In practice, the experts often used the international forum as a leverage to realize local ambitions – a phenomenon the political scientists Margaret Keck and Kathryn Sikkink have described as the 'boomerang pattern'.[42] Pieter Gerbrand van Tienhoven – a Dutch ornithologist, insurance agent and international networker for conservation – skillfully exploited this strategy. 'A voice from abroad', so he stressed in a letter to a friend, 'always resonates more powerfully with administrators than the voice of a compatriot'.[43]

It was in the mixed networks of men of power and men of science that the pre-war ideas of Sarasin would be worked out in detail. In line with Sarasin the scientific aura of the undertaking was continuously reaffirmed, if only to convince the public that the conservationists did not constitute a 'silly nature-hysterical movement'.[44] The focus on 'primitive' areas in the colonies remained as well, as did Sarasin's ideas about how to protect these. Scientists active in the Belgian Congo, for instance, upheld an ideal of 'biological orthodoxy', which implied that national parks should be shielded from every kind of human activity. It was not only a selection of species, but the 'natural equilibrium' as a whole that was to be protected.[45] In other contexts it was clear, however, that the utilitarian discourse that had typified the early ornithological conferences persisted as well. At the international conference of London in 1933, Sir William Gowers pleaded to actively destroy 'vermin' through the use of traps and poison. He was (still) targeting 'destructive' species such as lions, leopards and hyenas. The zoologists present did not voice opposition.[46]

The tropics (and Africa in particular) might have been at the heart of the conservationist attention, but Sarasin's legacy did not limit itself to the colonies. With some rhetorical skill European nature could also be framed as an object of global concern. At the 1931 conference in Paris, for instance, the zoologist Clement Bressou stressed the international importance of the Camargue region in southern France. This, he claimed, resided in the fact that it was an important stopping place for migratory birds, that it was one of the last sanctuaries of certain, previously 'international' species (such as the European beaver), and that it offered a 'life community' that would help the international scientific community in solving the great biological mysteries.[47] At the same conference two Spanish speakers defended the international relevance of the nature in their country with the argument that Spain had historically served as a bridge for the spread of northern species to the south, and vice versa. At least at a rhetorical level, experts proved willing to look beyond their national borders.[48]

New Expertise, Old Institutions

The discourse of the internationality of nature protection offered the possibility to reinvent old practices and re-establish institutional reputations. Natural history museums offer a good illustration of this. As centres of biological research they had lost part of their prestige to university laboratories in the late nineteenth century, but international nature protection gave a new sense of urgency to their project of globally inventorying species.[49] From the 1920s onward, curators of European natural history museums increasingly profiled themselves as specialists with regard to international nature conservation. These curators both played the role of activist, pressuring governments to act, and the role of expert, offering the same governments purportedly neutral advice.[50]

A good example is offered by Abel Gruvel, Professor at the Muséum d'Histoire Naturelle in Paris, and the most visible French zoologist engaged in international nature protection. At the Conference of Paris in 1923 he issued a recommendation that called for the protection of the fauna in French colonial Africa. As a direct result the French Ministry of Colonies contacted Gruvel himself to make up a list of species and places to be protected. Gruvel drew up the list, and did not fail to indicate that the exploration of the national parks should be a work of *French* scientists. Exceptions to that rule could only be accepted after consultation with his Muséum d'Histoire Naturelle. The international arena had thus helped to further the cause of one national protection movement, the power of one scientific institution and the career of one man.[51]

As natural history museums had, zoological gardens also jumped on the wagon of nature protection. From places of popular entertainment and acclimatization of foreign species, they increasingly turned into centres of scientific nature conservation that engaged in the breeding of threatened animals.[52] Even when the species concerned could be localized to a specific habitat, these breeding programmes often involved zoos from a great variety of countries. As such they were international almost by nature. One of the most publicized examples of such an international breeding programme was that of the European bison. Its American counterpart had gained iconic status in the American conservation movement, because it had secured the survival of the species at the turn of the century.[53] The European bison, however, became extinct in the wild in 1927 – only five years after the foundation of the (mostly Polish-German) International Society for the Protection of the European bison. Thus, rather quickly, this society was engaged with a species that only survived in captivity.

Because of the geographical dispersion of the last surviving European bison, the society was necessarily international in outlook. This, again, did not obliterate national tensions. Poland, which had housed the last wild animals in the forest of Białowieża, and Germany, where most breeders were active, often clashed. The tensions are a returning theme in the letters that the director of the society, the Frankfurt Zoo director Kurt Priemel, addressed to his American patrons. He stressed that the Poles mostly propagated the European bison as a 'national animal' and showed reluctance for sending breeding animals abroad. Later he would add that Germany was more suited to concentrate large numbers of bison, because in Poland wars were to be expected that could harm the animals.[54] His American correspondent, the New York Zoo director W. Reid Blair, replied with irritation that he did not consider national politics an appropriate topic of conversation.[55]

The story of the European bison shows how thin the veneer of internationalism often was. This became even more clear in the following years. After the takeover of the National Socialists in 1933, the International Society was complemented with a strictly German committee of bison breeders. Its director was no one less than Hermann Göring, who saw the breeding programme as instrumental in recreating the pure animal world of a prehistoric Aryan past.[56] Several breeders (always short of money) engaged both in the national and the international project. Lutz Heck, the director of the Berlin Zoo, for instance, was both the brain behind Aryan fantasies and a proponent of international collaboration. When the German army entered Poland in 1939 he immediately wrote to his Dutch friend, Van Tienhoven, to reassure the latter that the Polish bison were doing fine.[57] He did not inform him, however, about the ethnical cleansings ordered by Göring to make Białowieża into a 'primitive' wilderness. Neither did he enlighten him about his own initiative to transport the only bison of the Warsaw zoo to Berlin, nor about the slaughtering he and a group of SS soldiers held among the other animals of the Polish zoological garden.[58] Zoos, thus, might have easily picked up on the international aura of nature conservation, but this does not mean they escaped the nationalist cruelties of the period.

Next to natural history museums and (partially) zoos, the aforementioned national parks were also integrated in the international conservation project. This went together with a transformation of their original outlook. The national park, an American invention of the 1870s, was originally hardly conceived as a scientifically managed reserve, but rather as a means to protect picturesque panoramas and to make best use of tourist opportunities.[59] In the 1920s and 1930s, however, the small transnational network of zoologists saw possibilities to turn national parks into places of science. Their most successful attempt to do so materialized in the Belgian Congo. Inspired by the work of the American taxidermist Carl Akeley, and initiated by the Belgian king, an expedition was sent out to prospect possible territories. The Belgian zoologist Jean-Marie Derscheid led the expedition that would delineate the actual boundaries of the eventual park, and became the first secretary general when the newly founded Albert National Park opened its doors in 1930. Under pressure from, amongst others, Derscheid, the park would be managed by a commission dominated by scientists. Of course, the national park acted as a source of national Belgian pride, but it was nonetheless inscribed in the rhetoric of international conservation. Not only did its zoologist-managers stress at every occasion that the park counted as a common heritage of mankind, but they also made sure to include foreign scientists in the managing commission, and to openly promote it abroad as a place of research.[60] Furthermore they would use international conferences to pressure the British to turn the bordering areas in Uganda into a national park as well – as such creating a transnational form of nature protection.[61]

Next to museum naturalists, zoo breeders and national park managers, the international protection movement of the interwar period produced one final type of expert: the clerk. The activities of conservation experts across borders asked for coordination and a good knowledge of the differing legal and ecological situations in the various countries. The aforementioned International Office for Documentation and Correlation of the Protection of Nature was set up to do exactly that. Its prime initiators were (again) Derscheid and Van Tienhoven. Its professed goal was 'to centralize, classify, publish and deliver to everyone who is interested the documents, legislative texts, information and communications of all kinds that relate to the protection of nature'.[62] By 1931, no less than 8,000 index cards had been filled in to process the collected data.[63] More important, however, was that the office also served as a symbol of expertise and as the centre in a network of correspondence. In a letter to a colleague the well-networked Van Tienhoven indicated the importance of the latter: 'Relations ... of prominent men in different countries are in my opinion the principal base through which we must influence our governments.'[64] Putting experts in the position to actively influence policy was the actual goal of the organization. The office could foster the old boys network, and bring scientists into contact with powerful men from the world of diplomacy, the colonial administration and the royal courts. It was a work of correlation indeed.

By creating new institutions and taking over old ones, conservation experts effectively managed to influence policymakers. In the interwar years, the international network of conservation experts literally drew borders of national parks (for instance in Belgian Congo) or shaped international hunting legislation (in the London conferences). This effectiveness was the result of four conditions. First, there was the experts' success in creating a coherent identity for themselves. Second, the experts managed to use this identity to foster a network for gathering information. Third, they succeeded in stressing the policy relevance of this information in their continuous informal contacts with policymakers. And, finally, the growing importance of the state in nature protection (through national park management or hunting legislation for instance) made that conservationists could translate their ideas into practice. In all these four aspects the world of around 1930 drastically differed from that around 1900.

The growing influence of zoological expertise was not achieved without hard work. It implied that zoologists actively engaged in networking, lobbying and administrative activities. Yet, the role of networker, lobbyist and administrator not necessarily combined well with zoological work – neither theoretically nor geographically. When Derscheid was in Africa doing fieldwork, Van Tienhoven wrote to Derscheid's wife, Jeanne, that 'it would be a pity if Jean-Marie would come in a position in which he is situated far from us and from the work we could do'.[65] With 'work', Van Tienhoven referred not to zoological, but 'correlation' work. The reply of Jeanne Derscheid is equally interesting. She wrote:

> Indeed, we believe that for the development of the great work of nature protection we need persons who devote themselves to it with body and soul, but it is equally important that these persons maintain their scientific titles in the world in which they perform their mission, and this in order to enforce their authority.[66]

The quote is revealing about how Miss Derscheid saw the relation between science and nature protection. Protection concerned Derscheid's work as a clerk and activist in Brussels; science was what was done in the national park. This scientific activity, the argument went, was necessary to speak with an authoritative voice in Brussels. Epistemic authority (acquired in the field of science) could, thus, be translated into moral authority (in the field of activism).

Kill, Protect, Control

The early nature protection movement mobilized naturalists affiliated to museums, zoos, national parks and documentation offices. Through these institutions, the scientists tried to get a grip on the worldwide management of wildlife, and to prevent the latter's destruction. This did not imply, however, that they would unambiguously distance themselves from the killing of wild animals. Practices of hunting, after all, were deeply embedded in the cultural milieu from which the nature protection movement originated. Hunting constituted a substantial aspect of the field ethos of the naturalists, and it was part and parcel of the culture of the colonial officers and aristocrats they took as their patrons. To be truthful, conservation-minded scientists criticized the killing of wild animals on an industrial scale, but at the same time many of them also defended more sustainable types of individual hunting. In this, they referred to codes of sportsmanship and masculinity. As such they (at least partially) echoed colonial and aristocratic cultural codes, in which the individual confrontation with wild animals was presented as a good exercise in character building.[67] This explains why Van Tienhoven, while he openly fulminated against hunting trips organized by the Royal Dutch-Indian Airline Company in French Indochina, also described a friend as a 'full-blooded hunter ... of the good sort'.[68] Who exactly belonged to this 'good sort' was a topic of discussion within the conservation movement. A case in point is the clash between Victor Van Straelen, the director of the Institute of National Parks of Belgian Congo, and Léon Lippens, amateur ornithologist and secretary of the International Committee for Bird Protection. Van Straelen repeatedly complained that Lippens sided too easily with 'the hunters and destructors', thus undermining the actual goal of his organization. For a while the conflict even threatened the subsistence of the shared secretariat of their respective organizations.[69]

As part of their ambition to create good hunting practices, the conservationists explicitly sought dialogue with hunting organizations. This explains, for instance, why representatives of the French Saint Hubert Club were invited as speakers to the Parisian nature protection conferences.[70] Likewise, delegates of conservation societies would lecture at hunting conferences. In their private letters, conservationists explicitly stressed the importance of infiltrating these congresses, since conservation was believed to be 'a matter that should not be left to the hunters alone, but that should be researched on a broader basis'.[71]

The entanglement of hunting culture and early conservation ethos was particularly clear at the two London conferences for the protection of the African flora and fauna. At the conference of 1933, it was the governor of Sierra Leone, Sir Arnold Hodgson, who chaired the expert discussion over which species should be granted protection. He took the forest hog from the list because it was 'not even recognized as a game animal'. The Beira antelope was struck off, because – according to the hunting experience of the discussants – it was protected well enough by the natural environment in which it lived. Other species, then, were explicitly put on the list to guarantee a continuation of hunting in the future. At the end of the discussion Hodgson proclaimed unequivocally that the conference did not want to get in the way of the 'true' hunter. The hunt, so he stressed, improved knowledge of nature, it kept men physically fit and it sharpened their taste for exploration. 'Big game hunters', he concluded, are 'a fine company of men whose cause I will always champion'. The zoologists present did not voice disapproval.[72]

Rather than to prohibit hunting, the early conservationists hoped to administer who was allowed to hunt and where they were allowed to do so. In line with this hope, they lobbied colonial administrators to interdict access to hunters whose behaviour was considered excessive. In this, a particular indignation was reserved for those hunters who claimed to 'collect scientifically', but who actually hunted rare animals 'purely for financial reasons'.[73] After all, not only the reputation of 'the true hunt' was at stake, but also that of 'true science'. Scientists were allowed to kill animals, rationally and in a restrained manner, but the 'false' scientists were to be unmasked.

What was held true for hunting, was, by extension, held true for the trade in living animals. Several of the early scientist-conservationists were involved in such trade in their capacity of zookeepers or as owners of private menageries.[74] As long as this trade took place in a scientific spirit, it was seen as part of conservation itself; if part of a commercial circuit, it should be actively and forcefully opposed. Yet, like in the case of big game hunting, the division line between right and wrong was not very straightforward – not even for the conservationists themselves. For this reason, Van Tienhoven tried to assemble lists of trustworthy and irresponsible traders of wild animals. He contacted members of his network in order to settle whether particular zoo directors had to be classified as 'a well-known savant or rather as a manager of a funfair'.[75] Again, it was the connection to science that served as the indicator to discern moral from immoral practices. Expertise was considered to produce knowledgeable, but, even more important, also honorable behavior.

Conclusion

The international conservation expert is a creation of the interwar years, but the context in which he was active has changed quite significantly since. To begin with, conservation institutions have seen a magnificent increase in scale. The postwar period witnessed the rise of gigantic expert-driven organizations such as the International Union for the Conservation of Nature or the World Wildlife Fund and the advent of monster conferences on conservation issues organized by the United Nations. The old boys network (although it never entirely disappeared) has become much more public-oriented. The experts have incorporated the language of new disciplines: ecology in the 1940s, conservation biology in the 1980s. And the rise of the catchwords 'environmental' around 1970 and of 'biodiversity' in the late 1980s, has also led to a new framing of conservational issues, both among scientists and the general public.[76]

Yet, in many ways the institutional and cultural infrastructure of the protection movement still resembles that of the interwar years. Still international conferences play a prominent role in the movement. For zoos and natural history museums the conservation of species has become even more important in their self-presentation. And national parks claim even more international relevance since they have been renamed 'biosphere reserves'. The main icons of protection have also remained the same. At least in the public eye whales, big African and Arctic mammals and migratory birds remain the main concerns of the movement. *Naturvölker* have been renamed 'indigenous people' and are now presented as 'stewards of the earth' rather than as objects of conservation, but organizations like the World Wildlife Fund still take up the defence of their cultural heritage.[77] In general, we might conclude that the constellation created in a particular context in the 1920s and 1930s still influences the debates today – despite the fact that the world has drastically changed.

The main idea that got hold in the interwar years was that nature protection should be international and science-based. Naturalists (and particularly zoologists) established themselves as the most visible experts of this international science-based conservation. Their expertise had not arisen as science on demand. It did not originate in a policy question raised by governments or administrations. It was rather the other way around. The zoologists started out as advocates, placing a problem on the agenda, and then trying to convince governments and administrations that they should be consulted about how to address it. In putting this message across, transnational networks proved crucial. These networks had their origin, amongst others, in international scientific conferences, which served as performative spaces. It was to a large extent there that the identity of the conservation expert was created and fostered. It was also largely there that the negotiations took place about which nature was (internationally) important, which particular knowledge was needed to protect it and which values were to be incarnated by its expert-protectors. Once the identity of the scientific conservation expert had been created, a phase of institutionalization followed. Pre-existing institutions such as museums, zoos and national parks were 'colonized', while new institutions such as nature protection offices were set up. They served both as tools and symbols of the conservation expert.

In the interwar years, a small group of people managed to spread the idea that wild nature was (at least in part) internationally valuable. The same people took up an expert role to make its protection possible. This expert role was based both on science and ethics. It concerned both knowledge claims and a particular ethos. Establishing the expert role, thus, involved boundary work – such as a strategic dissociation from 'unscientific hunters' or 'silly nature hysterics'. With this boundary work two audiences were targeted. To begin with, there was the inner group, who had to be convinced of its expert role and who had to internalize its values. Secondly, there were the men of power who had to be convinced their expertise was serious and of practical consequence. Only by addressing these two audiences, the conservation experts could – to repeat the words of Miss Derscheid– 'enforce their authority'.[78]

4 THE HOUR OF THE EXPERTS?: REFLECTIONS ON THE RISE OF EXPERTS IN INTERBELLUM EUROPE

Martin Kohlrausch

In 1944, at the height of the Second World War, Friedrich Hayek's *The Road to Serfdom* came out. One of the classic studies of economic writing in the twentieth century, it is for many a notorious, for others brilliant example of what later became the Chicago-school's interpretation of the world. Hayek's work stages a fervent refusal of what he calls, in inverted commas, the 'inevitability of Planning'.[1] In its core, however, his work is a study about the role of knowledge, which, Hayek believes, planning will always fall short to represent in all its complexities. Given that, as Hayek explained one year later in an equally classic article, 'scientific knowledge, occupies now so prominent a place in public imagination', it comes as no surprise that also those commanding this knowledge – experts or specialists as he calls them – are centre-stage in Hayek's account.[2]

The point Hayek makes in the *Road to Serfdom* is to deny a need for planning, stemming from technological progress, but interpreting planning as resulting from deliberate action, brought about by technical experts. What is more, for him this is a necessary outcome of the formation of modern experts itself:

> The movement for planning owes its present strength largely to the fact that, while planning is in the main still an ambition, it unites almost all the single-minded idealists, all the men and women who have devoted their lives to a single task.[3]

Hayek's experts are highly ambivalent figures: they are the pacesetters on the 'road to serfdom'. His insights could, Hayek claimed, reveal the truly dangerous movers and shakers. Yet, at the same time they are key figures for progress, as he has to acknowledge.

Hayek was an early and important voice hinting at the unaccountable and unchecked power of technical and scientific experts. More important in this context, his observation rightly points to the logic of the expert – beyond her or his very expertise – and the need to ground this logic historically. Beyond the point against planning, Hayek makes an interesting shift from the logic of planning to the logic of those who execute and bring about planning – a shift, in which historiography was not always quick to follow Hayek. Long before this became standard in respective scholarship, he referred to the 'Zeitgebundenheit' (the embeddedness in the experiences of one's time) of expertise and in this also to the relation of power and expertise – the politics of expertise. Without using these terms Hayek thus also hints at the performative and constructed aspects in the rise of modern experts.

When *The Road to Serfdom* came out, state-centred planning was at its height, bringing about seemingly spectacular results in Albert Speer's 'Zentrale Planung' (Central Planning Board) and of course in the Soviet Union, soon to dwarf Speer's overstated output figures.[4] Moreover, with the New Deal, the war economy in the US in full swing and John Maynard Keynes along with Lord Beveridge turning the tide towards planning in Britain, the phenomenon extended also to the western liberal market economies.[5] This 'spill over' from continental ideas to Britain was what finally alerted Hayek. Hayek had first-hand knowledge of the respective debates on the other side of the channel. Deliberately chosen as an antidote to Cambridge-based Keynes he had left his native Vienna and joined the London School of Economics in 1931. His insistence on the negative importance of experts drew on his experiences in Austria and on his reading of what happened in Germany in the years from 1933 on. Obviously, Hayek understood his interventions at least as much as political as economic. Even more than with planning itself (in his eyes more of a symptom) he was concerned with a misguided apprehension of science and the embellished role it ought to play in society and economy. Such views he detected in a number of influential contemporaries, among them his well-known compatriot, Otto Neurath, inventor of the ISOTYPE (International System of Typographic Picture Education), eloquent champion of scientific planning and, as political émigré, since 1940 also in London.[6]

Thus, Hayek was in a twofold way part of the changing role of experts: personally, having left behind a central Europe ever more haunted by brown and red visions of a planned society, and professionally as an advocate and analyst of a liberal economy. This article will take up both perspectives, which, of course, cannot be separated completely: first, the turning of the tide from the democratic 1920s to the authoritarian 1930s in central Europe. Second, Hayek's diagnosis, assuming that he was at least correct in detecting the 'hour of the expert' in this particular period. This means this article will ask how far the two decades after the First World War formed a transition period in the relation of experts and politics. The essay argues that, in any case, this was a period of rapidly increasing mutual dependency between both experts and politics.[7] After placing the questions asked here in the historiography on experts, the article will look at those regions, where the transformation of statehood highlights the new positions of experts particularly, and what this meant for the staging, the performance of expertise. Building on this, and using the example of urbanism, the article will then look at the new public relevance of experts being able to translate the relevance of the problems they dealt with into visibility.

Changing the Perspective on Experts: Experts as Objects of Accelerating Historical Change

The historically informed study of experts, as this volume also shows, is a quickly rising field. These studies more and more interpret experts in their historical context, against the background of their time and take account of the changing cultural and social conditions of science as has been rightly demanded by scholars such as Dominique Pestre.[8] What has been less done is a more general, historically informed interpretation of experts, in terms of abstraction above the many specialized studies and below the far more universal theories the sociology of experts provides. These studies would answer questions such as how far specific formations of experts – beyond what was induced by technological or economic change – emerge in a specific time-frame? How did their emergence relate to historical caesuras and breaks? To what degree were these European or even global developments?

The more abstract approaches, of course deliberately, attempt to overcome those factors specific to experts in time and space. In contrast, a historically informed account could start with a fresh view on the chronology of experts, as this volume also does with looking for the specifics of modern expertise vs. early modern expertise. As with all chronologies its purpose would not in the first place be to establish definite dates but rather to make us think how to structure and communicate the manifold findings and trends in the historiography of experts. Almost unquestioned, it is often assumed that the rise of experts followed its own paths, was to an important degree universal, and defined by a technoscientific logic. This has led many scholars to question the relevance of established caesuras – the World Wars, revolutions, foundations of states – for the history of expertise and experts in favour of narratives dictated by the rise of the knowledge society, or at least a more general history of knowledge. In looking at 'Science and Social Space' in historical perspective Margit Szöllösi-Janze unfolded the dynamics of institutionalized science in their relation to the political developments in Germany – e.g. the rise of National Socialism.[9] Szöllösi-Janze shows the relevance of long-term trends, which took their dynamics from inner-scientific developments but interacted intensely with the sphere of politics and society. Seminal events like the First World War should thus not be seen as breaks but rather catalysts, bringing about qualitative change. Indeed, the First World War triggered completely new demands on the state with ensuing opportunities for experts – and a growing mutual dependency of the expanding state and experts increasingly relying on the state as employer, regulator, etc. The post-war crisis, diagnosed and defined partially by experts themselves, became the framework for the self-empowerment of experts.[10] In an influential article on what he labelled the 'Scientization of the Social', Lutz Raphael substituted established caesuras and identified four phases of the growing importance of social expertise from the mid-nineteenth century to recent times.[11] The new experts of a modern kind were far more than just the products of professionalization. They were part and parcel of the process that was aptly described as the establishment of 'territoriality'; that is, the process of new forms of political control facilitated by technological progress, which, as historian Charles Maier argues, started in the mid-nineteenth century.[12]

Answering the demands of an ever more complex society, social experts could establish new fields – and career opportunities. They made use of the increased clout of science, often in an instrumental way. No doubt the changes Raphael describes were triggered not only by new scientific insights but also by social challenges and political frameworks, with in particular the First World War again seen as a catalyst. Raphael, as much as Szöllösi-Janze, integrates the upheavals of the age of extremes. Both are, however, operating from the perspective of rather anonymous processes in which experts certainly do not figure as mere executors of broader change, but also not as personae in their own right. We learn more about how experts shaped social and political change than what social and political change did to experts. Bringing back in the caesura might serve as a means to do both – to integrate a more personal perspective of the historical experience of the extreme breaks of the twentieth century in particular and – in linking these two developments – to come up with arguments pointing beyond individual experience but also to some extent beyond specific groups or regions.

In order to understand the impact of such caesuras, we have to recall the manifold ambivalences inherent in the expert status – some more general, some more specific to the first half of the twentieth century. Here I will particularly discuss the break of 1918. While experts strove to present an 'ideology of professionalism'[13] that primarily served the public good and not the state, the practice differed considerably. Situated in between the layman and the political decision-makers, experts cannot exist without the authority bestowed upon them by those in the position to make far-reaching decisions. On the other hand, in order to secure their position against the authorities, these experts relied on the empowerment of society. The question of the authority of expertise was linked in a rather direct way to political or social authorities.[14] Experts empowered themselves in new fields of political or social action. They relied heavily on a process of specialization, which gained momentum in the period under question, but at the same time tried to enter into new fields of action that were often only loosely connected to their original training. While professionalism was an important anchor of the self-perception of experts, public standing increasingly turned into a necessary additional feature.[15] Next to the state's growing dependence on experts, public standing came to form an essential source of autonomy. Yet, the more the state relied on experts, the stronger the political imperative to control experts became.

On the one hand, experts often forwarded – in a wide sense – democratic goals when spreading knowledge or furthering social progress. On the other hand, their striving for long-term planning and the putting through of large-scale projects often made them prone to follow authoritarian political concepts. Equally complex was the interplay between national and international sources of expert authority. While experts mostly defined their authority in a national framework, highlighting the universal grounding of their expertise was crucial to the experts' claims for influence.[16] Moreover, international contacts were an important criterion in attaining expert status. This tension often resulted in structural conflict between national loyalties and a universalistic self-understanding, which was typical of European experts, particularly in the twentieth century.[17]

How fragile the very distinction between national and transnational was, becomes evident when looking at the shifting geographic frameworks in which experts acted. This is particularly obvious in the empires, which dominated Europe well into the twentieth century: the three (or four if one includes the Ottoman Empire) land-based empires in the centre and in the east, and the colonial empires of France and Britain in the west. These empires were empires of opportunities for the new kind of experts that emerged in the nineteenth century: the infrastructures that held the empires together, along with the large new technological projects– the dams and traffic arteries that formed their new resources of power. In particular, the multinational empires and the colonial empires formed attractive structures by combining vast regions under one legal and political structure, leading to surprising degrees of mobility for those committed to solving the empires' technological challenges.[18] In central and eastern Europe these structures ceased in 1917/18, with experts, if they had not already done so in the preceding years, forced to realign their loyalties.

The relevance of the change in perspective towards the personal perspective of experts becomes particularly obvious if we turn to this part of the continent, which saw radical and accelerated change in the immediate years after the First World War, and if we ask what this meant for the performance of expertise.

Experts in the Postimperial Age

The break of 1918 entailed new demands on the legitimacy of the regimes which won the war – and had to make good for the sacrifices asked from their people.[19] This was far more pronounced in the centre and east of Europe. In the capitals of the new postmonarchic, postimperial states, from Petrograd/Moscow to Vienna, Budapest and Berlin, governments had to prove that they could outperform the empires they succeeded. Following the demobilization after the First World War, these states had to rebuild structures in what, in comparison with western Europe, sometimes came close to a vacuum.[20] Significantly, not only traffic arteries and administrative elites lacked in new states like Poland or the Baltic states, but also political legitimacy was in short supply.

The extreme need for legitimacy of the state in fields such as health and housing translated into new chances and challenges for the respective experts in these areas.[21] 'National engineering' offered enormous opportunities, but those who rose to new eminence also had to realize that it became more and more necessary to decide where their loyalties lay, beyond their mere professional identity. The flip side of this process was increasing constraints and coercion for technical experts who aligned themselves more closely than before with certain political regimes.[22]

With the establishment of new states, or at least new political systems like in Germany and Austria, and also to some extent in the victorious countries, technical experts were clearly in a critical position in those fields where these states needed to prove their ability to solve problems. This was even more so where military defeat and social crisis coincided or where the war, even with results perceived positively, had changed the political regime and brought about completely new political entities. In an extreme way, the break of 1918 in central Europe highlights a development that dates further back – the engagement of experts in the nation-building process.

Just before the outbreak of the First World War, Gyula Hevesi, the 'mastermind' of the mobilization of Hungarian engineers during the war, had complained:

> There is a much greater problem here to be solved, a problem of which my individual grievance is but a tiny, though inseparable, part. The task is to destroy the barricades that have been raised by today's society in the way of all who possess nothing but their talent and knowledge.[23]

An ardent reader and follower of Austrian economist Joseph Schumpeter, Hevesi believed that changes in know-how were to become the decisive process and factor of his time.

This was not only a strong symptom of a growing 'engineer consciousness'. It was also a stepping stone for a fervent anti-capitalist, state-oriented ideology that was to guide the majority of Hungarian engineers after the war that their country had lost so dramatically. Against the background of a drop in salaries that was sharp and disproportionately dramatic compared to other heavily affected groups, radical solutions became more attractive not only for Hevesi. Like engineers in other European countries, Hungarian engineers lamented what they perceived as their being sidelined in the process of capitalist production. They also believed that a scientific approach towards politics would harmonize with a scientifically grounded state socialism.[24] Protagonists of such an 'engineers' utopia' closely followed Russia's war socialism, but also the theoretical debate in the USA associated with the name Thorstein Veblen.[25]

What was specific about the situation in Hungary after the First World War was that a large proportion of engineers opted for political engagement and believed strongly, at least for some time, that they were close to the coordinating points of politics. After all, as these engineers informed all the 'technical science workers' in and outside Hungary via telegraph, there was a chance 'that the technical sciences in their purity will play the leading role in production'.[26] In fact, the outcome was much more prosaic, with bureaucracy rather than engineers initially assuming the political power.

Still, the example indicates that the challenges for the state posed by the Great War generally enhanced the position of the technical elites. In the newly created Polish state this link was most obvious as different tendencies came together here. The political system changed from a monarchic system to a republican system. Poles became the new dominant group with an ensuing change of elites, and the disappearance of Russian and German elites. Finally, an ambitious merger of three formerly completely separated economic territories from the German, Russian and Austro-Hungarian empires had to be achieved against the background of massive destructions. What for contemporaries was a most challenging situation formed an opportunity structure for experts, who not just profited from an ongoing scientification, or the need to cure the ills of modern society, but from a very concrete, politically induced need for legitimacy.

Also in Poland, engineer associations voiced demands for a stronger political say. During the First World War, Polish engineers effectively aligned their cause with that of the nation. These engineers believed that their expertise would also enable them to deal with social problems that were essential for the future of the nation. Hence engineers should occupy key positions in the higher administration. This led to the view that the training of engineers should receive particular attention, not just in terms of technical knowledge, but also with a regard to the skills necessary for the formation of a democratic society.[27] In Czechoslovakia, equally a state coming into being after the war, engineer Albín Bašus proposed a more integral technical education, which was to produce 'organizers' and 'leaders'– the technical experts needed to shape a modern nation.[28]

Such notions had a clear professional edge; for example, in demanding that no foreign specialists should take qualified positions in the administration that could be filled by Polish experts, or trying to prevent state intervention in what was perceived as the deserved rights of one's group. However, the reasoning and political agitation went far beyond such issues. Continuing after the war, the professional organizations of engineers argued that only the expertise of engineers would be able to overcome the social differences of society and, in particular, the problems of party politics. In line with technocratic conceptions prevailing all over Europe, Hungarian or Polish engineers claimed to be able to offer a neutral force around which effective government could be centred. The electricification of Poland, for example, was a project that extended far beyond simply providing energy for industry, but also touching on such crucial matters as hygiene and health, understood in a broad sense.[29] Technocratic models that were popular with Polish engineers had an extremely strong influence in Poland and also transformed politics more visibly than elsewhere. This was true for the whole notion of the *Sanacja* (healing) regime, established in 1926, but also the eminent role of 'state technicians', such as Eugeniusz Kwiatkowski, who oversaw huge planning schemes in the 1930s.[30]

The relevance assigned to these experts in the cause of nation building also had to do with their scarcity, which was a direct result of the pre-1918 period. A 1931-estimate counted only 25,000 technicians and engineers in a country of some 32 million inhabitants due to the restrictive policy of the partition powers lasting until the First World War. Only about 10,000 of these had graduated.[31] Against this background, experts were actively encouraged to return from western Europe and the US.[32]

The Challenge of Serving the Public: The Example of Urbanism

These developments seem, at first glance, a result of a very specific situation in an area of deep transformation of political and administrative structures. Yet, in many ways what was going on in Hungary or Poland reflects more general European developments. In Germany, which saw a radical political transformation from November 1918 on, the terms *Leistungsverwaltung* and *Daseinsvorsorge*, which could loosely be translated as 'public service', came up in the 1920s and were grounded scientifically in the late 1930s by law professor Ernst Forsthoff.[33] Forsthoff, later notorious for his closeness to the Nazi regime, managed to aptly grasp a new dimension of state activity. The state, in Forsthoff's eyes, had to react to the fact that the individual was on the one hand confined to an ever smaller circle of living, while on the other hand commanding ever more opportunities to reach out via technological means. In this the individual became dependent on state-sponsored services. As this was a structural process, the state had to intervene.

Beyond its juridical merits these terms convey well that the state's extended reach of activity formed a chance but also a challenge. Looking at the challenges posed for the state is decisive for the post-1918 situation. Both the legitimacy and, indirectly, also the sovereignty of states increasingly depended on expert knowledge. The Great War had demonstrated this connection in a dramatic way, but it was in peacetime when experts themselves – as Hayek captured negatively – could define the limits of expert-based activity. Whole new areas of expert activity emerged out of the tension between new technological opportunities and new demands on state activity.[34] Historian of science Caspar Hirschi's argument that the modern societies of the last two hundred years distinguished themselves from earlier periods by the 'invention of innovation' seems plausible. Hirschi correctly stresses not only the new speed of innovation, but also that innovation was 'embraced as an imperative practice'. Innovation attained a legitimizing function.[35]

After the First World War, as philosopher Tzvetan Todorov rightly put it, science

> was no longer content to restrict itself to acquiring knowledge of the world but sought instead to transform the world, in keeping with Marx's tenet, in order to achieve an ideal that, it claimed, was rigorously deduced from scientific observation.[36]

The striking rise of urbanism after 1918 provides a particularly telling example. The emergence of urbanism followed both new supplies – of planning knowledge – and new demands on the side of the state to integrate an urban space which was perceived as more complex and more problematic, and in this as a task for state intervention. Yet, the rise of urbanism also resulted from the ability of a new class of experts to widen the frame of what was needed and could be solved (and those potentially able to solve these problems). Urbanist-architect Kurt Schumacher claimed that the architect had, by legitimizing himself, intervened into questions which per se are beyond his artistic sphere. The architect's thinking, Schumacher concluded, widened and he intervened in the 'territory of social problems, economic problems and technical problems'.[37] Adolf Behne believed equally in the interwar period that the 'architect is today easily more hygienic than the hygienist, more sociological than the sociologist, more statistical than the statistician and biological than the biologist'.[38]

The most striking embodiment of the urbanist strive to regulate large parts of daily life in a universal way may be found in the Congrès International d'Architecture Moderne (CIAM). The CIAM also provides a telling example of the relation of experts and politics, but also political ideologies as an important characteristic of modern expertise. The CIAM luminaries around architect Le Corbusier would not limit their political ambitions to building better houses, but envisioned whole new societies, at the same time claiming that no one else would be in so good a position to do so, as these were commanders of a new kind of meta-expertise.

In August 1933, Le Corbusier used dramatic language to underline the urgency of additional action, in this case a resolution clarifying CIAM's goals to Sigfried Giedion, the organization's mastermind and secretary general. Le Corbusier evoked those in executive positions – ministers and other politicians – as waiting for a statement of the CIAM and its practical action. 'It is high time, Giedion', said Le Corbusier, 'the world is on fire. There is a need for reinforcement. We are the technicians of modern architecture. In the name of due procedure and of the holy cause I demand that the resolution will be published'.[39] The CIAM experts dreamed of penetrating all areas of life with new progressive solutions derived from technological progress. They linked technology to a social cause.

Few other fields embodied this expertise with a cause in the way that urbanism did. This rather young discipline obtained its legitimacy by incorporating what were referred to as 'scientific methods', which dwelt on the extreme credibility that science and technology had acquired in the previous decades.[40] Since 1900, methods like statistical comparison and the use of newly available visual material like aerial photography, surveys and sociological analyses contributed to the idea that a planned development of all aspects of the city – if not of society as a whole – was not only desirable, but also achievable. Urban planning, however, particularly on the continent, often carried more expectations than simply improving the organization of a city.[41] In tackling the ills of the modern city, it strove to tackle the ills of modernity itself. Radical urban planners envisioned a new society and the rise of the 'new man'. This surplus of expectations was also a reaction to new technological possibilities, whether real or imagined. Almost necessarily, urban planners became technoscientific experts with a close relation to the state and society. They were strongly tied to the political, social and cultural developments and debates of their time.

Architects like Walter Gropius or Le Corbusier, both active in CIAM, as well as Giedion and many other lesser known CIAM members, very actively contributed to shaping the imagination to which they themselves added very concrete contributions. One may think of the surprising scope of action Bauhaus-radicals had in building, social transformation and education (at least for a certain period) or the impressive Red-Vienna – projects equally inspired by the promises of scientific urbanism and the political dividends it was hoped to distribute.[42] Here experts put their stakes in the great visions, which they themselves decisively helped to shape, and which were essentially bound to the interwar period.

The New Relevance of Experts and their Public Visibility

From the mid-1920s on, Giedion became one of the foremost 'translators' between the general public and the technical sphere, and promoter of new expert groups– a paradigmatic techno-intellectual of a new kind who helped to widen the frame of what was perceived as territory for expert action.[43] If we look at modern architects in the interwar period, what could be called writing architects are a striking novelty. Arguably, Le Corbusier or Walter Gropius made their careers less through their proven building talent (with very few examples in their early careers), or formal qualifications (which they both lacked), than their tapping on the hottest discourses of the day and making brilliant use of the newest media. Of course, we know since the works of Michel Callon on the importance of 'traduction' how important it is for experts to frame a problem, to make a case.[44] Yet, in particular for this reason it is worthwhile to analyze the conditions for the possibility to do so in a historical perspective. Gropius, Le Corbusier and others transformed their image into trademarks. In their cases it was often more important who said something than what it exactly was. Personality and charisma alone would hardly explain their rise to public fame. Rather the combination of embodying technological progress, a talent to find bold expressions for the new chances, and also their capacities and talent to lead exceptionally modern lives themselves may serve as an explanation.[45] An idea of this combination may be vividly grasped in László Moholy-Nagy's film *Architects' Congress*, shot at the CIAM IV congress and attesting to the self-perception of these architects as harbingers of modernity.[46]

Successfully the CIAM luminaries projected themselves as personifications of the social imaginary. They tapped the utopian space, opening up against the background of the experience of enormous technological progress and a deep political and social crisis. Yet, they were also able to grasp the internationalism which economic and technological exchange incurred. This development is even more noteworthy, as the idea that experts had to stay away from the worldly temptations of mediated celebrity was part and parcel of self-conceptions of neutral expertise and impartial science. Experts in many fields had an ethos of working behind the scenes. In some instances this was also tactically motivated. In any case, public standing and professional relevance – however measured – do not necessarily coincide.

There was, however, an important precondition, essential for these visionaries, which has not been mentioned yet. The emergence of new media changed also, among many other things, the scope of experts, at least in some selected fields. The innovations in offset-printing after 1900, which made the spreading of high-quality images much easier, as much as the rise of cinema, helped those trades from modernist architecture to early rocket engineering, which had visually potent material on offer.[47] They could enter a public space, partially nationally defined, partially internationally opened up by the new media. The symbolic dimension of technology certainly helped here, in particular when it fit into the mechanisms of mass media reporting.[48] Yet, men like Le Corbusier did not enter newspaper cover only because of the glamour of their projects, but also because of the vision which came along – to cure the ills of the past and modernity alike with the means of modernity. Unlike the inventors and scientists who came to prominence in the nineteenth century and served as the public pride of the nation, Le Corbusier and Gropius, highly controversial in their home countries, were hailed as problem solvers with a potentially global reach.[49] The visionary potential inherent in urbanism reflected back on the experts who personified this potential. Their fame as technocelebrities of the period between the mid 1920s and early 1960s could be understood as an advancement on the seemingly promising future in which social and political tensions would be eased or even wiped out by technological advancement and new planning insights. Men like Le Corbusier also attest to the remarkable personalization of technological progress and the expectations attached to it. This went far beyond the rise of a new technocratic elite of engineers and could even be found in the generally more group-orientated cult around engineers in the communist part of Europe.[50]

Stanisław Brukalski, one of the leading architects and thinkers of interwar modernism in Poland and also internationally, reflected this link well in 1935. Brukalski argued that against the background of impressive technological progress, far-reaching visions of new ideal cities were no longer a utopia and their planning not the occupation of 'scholarly madmen'. The challenge for experts, Brukalski claimed, was now rather to win over the public for one's visions and insights. The successful expert was a propaganda expert.[51]

The public dimension of experts and expertise coming to the fore here is historically contingent and thus subject to historical change: for technical reasons – the innovations just mentioned – but also for the powerful interplay between new media and a new kind of expert celebrities speaking to the particular challenges of the time. Also in this respect the interwar period formed an important transition period. The media-experts were exceptions in comparison with far more numerous others who stood not in the limelight. Yet, these media experts attest to the new clout, and also to new degrees of autonomy experts could gain after the Great War.

The construction of expert celebrities is a moment where we may tap into newly developed expectations on experts. Media interest may serve as an indicator, but in helping to bolster expert influence it also became an important historical factor in its own right. If experts are a social phenomenon, then the different way of talking about experts may claim some relevance. It seems plausible to assume qualitative change in the mid-nineteenth century. Now, experts had new means at their disposal – international reputation, a status as harbingers of national interest combined with social progress. All of this depended largely on new media and intensified communication that was not in place some decades before. These public experts were decisively transnational figures, in many respects building their status on the very fact that they attained a standing in different national contexts, using international relevance to change one's position at home and vice versa. On the other hand – and this is a somewhat better researched phenomenon – experts attained a status as national heroes, often going hand in hand with a public status confined to the very country of origin.

Conclusion

This article started with the observation of economist Friedrich von Hayek in 1941 that this was a time of unleashed power of experts. Against the background of this remark the article recalled the insights of recent historiography of experts, which understands the rise of experts as following the rise of technology and science and, increasingly in the twentieth century, the new social demands in ever more complex societies. While such readings argue – generally convincingly – from the intrinsic development of experts in their social context (thus transcending historical caesuras) this article stresses the importance of historical breaks, in particular of 1918. With varying degrees depending on the different regions in Europe, sharp political, economic and social breaks created voids of legitimacy which again translated into specific chances for experts. This was particularly expressed in central and eastern Europe, but generally in the whole of Europe we may see experts rise to new relevance in the interwar period. When stressing the ability of experts to answer such new needs, it is important not only to refer to the latter's capacity of self-empowerment, but also to look at concrete historical changes which in some ways formed preconditions for self-empowerment.

Highlighting the interwar period as a decisive period for re-framing the scope of experts raises of course questions of continuity. Obviously the kind of expert-politics connection, as expressed in the radical totalitarian projects like the *Generalplan-Ost* (Master Plan East) or the far-reaching transformations planned in the Soviet Union, caused a backlash and led to a hedging and stricter democratic control of experts in post-war western Europe. While in western Europe it seems the old, mainly law-based, administrative elites quickly controlled the power resources again, in the east the new technical experts, who became so important also politically in the inter-war period, appear to have played a longer lasting role in the new planning bodies emerging in the 1930s. Again, this might be a reminder to integrate carefully the breaks of the twentieth century into a more systematic history of experts in twentieth century Europe, still to be written. Looking at the public role of experts in its historical contingency may offer, as I argued here, a meaningful perspective to better place experts in the historical changes of their time.

When Hayek identified experts as men of the hour in 1944, this reflected both perspectives discussed here. Experts became critical not only because of the growing inherent relevance of the fields they represented, but also because political legitimacy drew much more heavily on such experts. These idealists proposed causes which did not necessarily have an inherent relevance, but were shaped by intense public discussions, hence also often by experts turned public figures. This is more visible with some groups than with others. After 1918 architects, in deciding for a specific way to built, also expressed or declined their loyalty to a new political system and its vision. Their insistence on the laws of science, technology and progress was far from neutral, at least far from being perceived as neutral. Also, and necessarily so, the sheer complexity and size of the projects that 'experts-gone-social' claimed and were attributed brought them into the vicinity of the authoritarian and totalitarian regimes. Already before being forced to decide between the totalitarian regimes, a number of western architects moved to the Soviet Union in the hope of getting there the means, but also the political backing, to put through their grand schemes.[52]

Hayek himself might serve as an example of how the age of extremes shaped a particular type of expert. Though certainly not in a dramatic way, Hayek's escape to London reflects this. His experience in Austria, where expert-based social improvement was meant to overcome the trauma of 1918, informed his negative view of experts. Tellingly, Hayek was himself much closer to the idealistic experts he despised than he would have it. Although he always advocated staying out of politics, when masterminding the neo-liberal Mount Pelerin Society in 1947 Hayek pressed his brothers-in-arms to come up with a utopian vision of liberalism to counter the lure of the left. It was not without bitter irony that once Hayek was eventually acknowledged as a leading economist with the Nobel Prize in economics in 1975, his fellow prize-winner was no other than Swedish Gunnar Myrdal, arguably the very embodiment of the idealistic expert in the twentieth century.

5 THE PSYCHIATRIST AS THE LEADER OF THE NATION: PSYCHO-POLITICAL EXPERTISE AFTER THE GERMAN REVOLUTION, 1918–19

David Freis

The history of psychiatry and its diagnoses has always been a history of social and political norms, beliefs and imaginations. This observation, now somewhat of a truism in the historiography of psychiatry, is especially true in the case of a particular sort of writings in which the diagnostic tools of psychiatry were not used to understand individual pathologies, but rather to explain the condition of society and the causes of political events. This theme can be found both in the professional discourse of medicine and psychiatry, as well as in political journalism from the early nineteenth century onwards. Volker Roelcke has argued that these writings should be understood as a medium of a 'bourgeois interpretation of the world and the self' and as a reaction to the crisis of bourgeois self-perception in the long nineteenth century.[1] However, the 'psycho-political' diagnoses can also be considered as a noteworthy example of a performance of scientific expertise in a socio-political context. As I will show with the example of the period immediately after the First World War, the diagnosis of social and political events as symptoms of psychopathological processes entailed psychiatrists' claim for an extension of their medical expertise onto socio-political matters.

The aftermath of the First World War saw a dramatic surge in the socio-political writings of psychiatrists in Germany and Austria. Shortly after the military defeat and the revolution, leading psychiatrists resorted to the concepts of their discipline to diagnose the current events as the work of anti-social 'psychopaths' and as the result of a national nervous breakdown and collective hysteria.[2] While intellectual and political elites imagined the national body (*Volkskörper*) as a sick body, psychiatrists offered a psychological version of this metaphor.[3] Many of the concepts used after 1918 had already been discussed in the nineteenth century, such as, for example, Cesare Lombroso's figure of the 'born criminal' or Gustave

Le Bon's crowd psychology (*psychologie des foules*). Yet, while these concepts had been the expression of the diffuse fears of an educated bourgeoisie towards the anonymous masses and criminal 'inferiors' in the late nineteenth century, they now seemed to offer an accurate scientific explanation for the very urgent socio-political crisis of the post-war period.

In a way typical for interwar cultural and political discourses, pessimism and optimism were inextricably linked with each other.[4] In a social and political situation in which many conservative psychiatrists saw the nation in a state of existential crisis, older concepts were not only updated and politicized but were also used to legitimize demands for far-reaching socio-medical interventions in the service of the rescue and regeneration of the national collective. Diagnosing the political crisis as a medical situation, and with a profound belief in the ability of the modern sciences to shape society, psychiatrists claimed for themselves the role of socio-political experts.

The politicization of psychiatric expertise in Germany and Austria was a consequence of the discipline's history before and during the First World War. Although it only became part of the medical curricula in 1901, psychiatry had successfully positioned itself close to core functions of the state, including the confinement of the insane and expert testimonies in the courtroom in the second half of the nineteenth century. At the same time, psychiatric concepts played an important role in the discourse on the bourgeois self. But unlike many fields of somatic medicine, psychiatry was unable to gain social prestige or financial resources from the contemporary breakthroughs in the laboratory sciences, like bacteriology and physiology. Despite important advances in the nosology of mental illness, psychiatry remained notoriously unable to effectively heal its patients. The First World War offered considerable chances for psychiatric experts to improve their position: the war marked a new height in the prestige and political relevance of the discipline.[5] The 'active treatment' of the so-called 'war neurotics' seemed to be a therapeutic breakthrough. It promised to end psychiatry's notorious inability to heal its patients and to usher an era of 'heroic therapies'. With the military stalemate along the Western Front, the mental health and resilience of the fighting troops and the nation were regarded as decisive strategic assets. After the defeat and the armistice, as well as against growing patient discontent and public protest against the often brutal methods of 'active treatment' and the allocation of veteran pensions, the discipline's wartime gains were in danger of unravelling. Psychiatrists' claim for expertise may well be understood as an attempt to defend and extend the wartime gains of their profession in the time following demobilization.[6]

Needless to say, this 'psychiatric need for expansion', as the 'anti-psychiatric' journal *Die Irrenrechts-Reform* put it in 1919, needs to be examined in the context of broader developments in the interwar period.[7] Psychiatrists' alarmist diagnosis of a collective 'nervous breakdown' and their warnings of an imminent collapse of German culture were also part of an ubiquitous discourse on the 'crisis' of the Weimar Republic and a general 'dramatization of the political imaginary'.[8] This alarmism mobilized and legitimized visions of national regeneration and a profound restructuring of society and politics with rational and scientific methods. Psychiatrists' aspirations for becoming socio-political experts led the discipline into a contested field, in which experts from other disciplines had already successfully staked their claims in the ongoing process of a 'scientization of the social'.[9] For example, Erwin Stransky, whose programme of 'applied psychiatry' was certainly the most radical attempt to extend psychiatry's expertise into all fields of social and political life, was convinced that psychiatrists' expert status had to be achieved not only against the anti-psychiatric bias of the 'public opinion' but also against the established expertise of jurists.

This chapter examines the socio-political writings of German and Austrian psychiatrists in the immediate aftermath of the First World War and the German Revolution of 1918/19. A first section focuses on the diagnosis of individual participants of the upheaval as 'psychopaths', showing how an already morally charged criminological concept became explicitly politicized in the wake of the revolution. A second section turns to the transfer of concepts of individual pathology to the national collective and examines how psychiatrists used the converging of clinical and political phenomena to cast themselves in the role of doctors of the nation. A third section will concisely discuss the further history of psychiatric expertise in the interwar period, and ask how and if these psychiatrists could actually capitalize on their psycho-political diagnoses of the revolution.

Throughout this chapter, the focus will be almost exclusively on the mainstream of German and Austrian psychiatry. Its protagonists were members of a conservative and educated bourgeoisie (*Bildungsbürgertum*), which was united by its common rejection of the revolution and the new republican order, as well as by its fierce nationalism. Although those psychiatrists, who published explicit socio-political diagnoses after the armistice and the revolution, were only a small minority of the professional group, their views were probably shared by a majority. This assumption is also underpinned by the fact that some of the psychiatrists who wrote about the psychopathology of political events were among the most prominent and renowned representatives of the discipline, such as the professors Robert Gaupp, Karl Bonhoeffer and Emil Kraepelin.

The Revolutionary Psychopaths

Eugen Kahn argued in the Munich medical weekly in August 1919:

> It has long been known that in times of turmoil, those prone to mentally disorders (*psychisch anfällige*) come forward, and after the experiences we psychiatrists have

made during the war, it did not come as a surprise to us that in the latest upheaval such people have stood in the fore'.[10]

In the short period following the end of the war and the revolution, a number of articles in different professional journals made an almost identical claim, pointing out that 'inferior' or 'psychopathic' individuals had been a driving force of the recent upheaval. As a member of Emil Kraepelin's psychiatric clinic in Munich, Kahn had been in a particularly good position when it came to examining the psychopathological dimension of the revolution. Munich had been one of the centres of the German Revolution and, for a short time in the spring of 1919, the capital of a Bavarian Soviet Republic. After loyal troops of the German army and right-wing *Freikorps* militia had violently crushed the revolution in May 1919, many of the survivors were imprisoned and thus came to be the objects of forensic examination by local psychiatrists such as Kahn, Kraepelin and Ernst Rüdin.[11]

On 3 August 1919, Kahn presented his findings at the yearly conference of Bavarian psychiatrists. He positioned himself as a scientific expert outside and above political struggles: to speak about recent events, Kahn argued, obviously held the danger of being caught up in the current political disputes. However, 'the idea that the psychiatrist always has to be ready to provide his judgment impartially and to the best of his knowledge, can and must help us to get over these concerns'.[12]

Although Kahn explicitly stated that he did not consider the revolution as such to be a pathological event and that not every revolutionary was necessarily 'mentally inferior', he had little doubt that 'psychopaths' had played a most important role during the upheaval. Of the sixty-six revolutionary leaders who constituted his sample, 'scarcely one could be seen as being overall mentally intact' and all the fifteen cases on which he reported in detail were to be considered as model types of the 'revolutionary psychopath'.[13] Among them were prominent leaders of the Munich Soviet, thinly disguised by pseudonyms: Otto Wasner (alias Kurt Eisner), Werner Leidig (Erich Mühsam) and Erwin Sinner (Ernst Toller).[14]

However, the concept of the 'psychopath' was far from being precisely defined and was mainly used as a description for a whole range of perceived 'abnormalities' in the grey area between normality and full-blown mental illness. As Kahn argued in the Munich medical weekly, the diagnosis mostly applied to personalities who would not appear as mentally ill, but nonetheless had mental deficits 'which lead them to make wrong life decisions often enough and to fail with them'.[15] The diagnosis of 'psychopathy' relied mostly on a necessarily normative assessment of a person's general decisions in life in terms of right and successful or not. For a conservative like Eugen Kahn, joining a socialist revolution obviously was a wrong choice. The notion of 'psychopathy' had replaced the older, and equally broad concept of 'mental inferiority' in the years following the First World War. But nonetheless, it was hardly sufficient for an encompassing, scientific description of deviant behaviour. In order to come to a more precise definition of the perceived abnormalities, numerous, sometimes rather arbitrary types of 'psychopaths' were introduced.[16]

In the case of the Munich revolutionists, Kahn did not exactly follow the influential classification introduced by his teacher Emil Kraepelin in 1903, but distinguished four basic types of 'psychopaths': 'ethically defective psychopaths', 'hysterical personalities', 'fanatic psychopaths', and 'manic depressives'.[17] Nonetheless, and despite all attempts of conceptual differentiation, the 'psychopath' remained a vague category, and once the light of 'psychopathy' fell on a person every aspect of his or her physiognomy or life could easily be read as a sign of abnormality. Against the background of wartime psychiatry and the growing criticism of its methods by patients and the public, it comes as no surprise that Kahn drew a direct line between the experiences of military psychiatrists and the revolution, claiming that the 'revolutionary psychopaths' belonged to the same group that had previously 'filled the military hospitals as war neurotics of all kinds' and had kept the military courts busy 'as elements that exceedingly threatened discipline'.[18]

Kahn was only one, but a largely representative example of a broader psychiatric discourse.[19] In the immediate post-war period, the assumption that 'psychopaths' had played an important, if not decisive role in the upheaval, seems to have been a largely undisputed consensus, as can be seen both in a number of articles explicitly dealing with the topic, and in casual remarks in many other publications.

When applying the diagnosis of 'psychopathy', psychiatrists used the conceptual framework of forensic psychiatry and criminal biology, asking for the pathological causes of individual deviant behaviour. In several cases, the examination of participants of the revolution took place in the context of criminal proceedings and thus at the contested interface of penal law and psychiatry, two institutions occupied with abnormal and deviant behaviour. However, more than by juridical motives, the use of the vague category of 'psychopathy' in the diagnosis of revolutionaries was driven by questions of politics and normalcy. Even when not used for political adversaries but for 'common criminals', the concept of 'psychopathy' was inherently political. Despite all attempts to introduce finely nuanced categories, and regardless of the replacement of the notion of 'inferiors' by the more scientific-sounding 'psychopaths' after the end of the war, the concept remained part of a morally charged 'dispositive of normality'.[20] In the grey area between madness and normality, the concept of 'psychopathy' allowed psychiatrists and criminal biologists to identify, construct and pathologize perceived threats against bourgeois society and morality.[21]

In the tumultuous situation of the immediate post-war period, 'psychopathy' offered the possibility to reframe perceived political threats as the object of medical and psychiatric expertise. Seeing the nation and the moral and political order of bourgeois society in peril and, at least in some cases, fearing for their own careers, positions and even lives, German psychiatrists eagerly used the propagandistic potentials of their diagnostic tools to discredit the revolution and its protagonists.[22] The diagnosis of 'psychopathy' for political adversaries offered the possibility to delegitimize their political claims by ascribing their actions not to any rational response to the current political situation, but rather to egoism, lust for power, the need to stand out, hysteria, or even to outright insanity. By shifting the analytical focus from the political to the clinical sphere, psychiatrists claimed for themselves the status of experts in a heated public debate.

As Paul Lerner has pointed out, this discourse must also be understood against the backdrop of military psychiatry during the war, and as a reaction to the public attacks against the practitioners of 'active treatment': by equating the revolutionists with their former patients, psychiatrists identified them as 'enemies of society' and as the ones responsible for the military defeat as well as for the violence and turmoil of the post-war period. In doing so, they not only denied their patients the status of victims but depicted them as the true perpetrators.[23] When the 'psychopaths' threatened society, German psychiatrists saw their duty not in the healing of the mentally ill but in defending society against them. Doris Kaufmann has claimed that this 'labelling and marking out of a group of so-called inferiors for their "failure in the war" has to be seen as highly significant for the scientific legitimation and acceptance of some practices of later national socialist population policy'.[24]

However, identifying the revolutionists as 'psychopaths' was not only a form of right-wing polemics or a way to make sense of a situation that seemed to challenge many of the certainties of pre-war society. By depicting society as being threatened by the 'psychopaths', psychiatrists positioned themselves in the first line of defence, claiming for themselves the status of socio-political experts. For the psychiatrist Hans Brennecke, the observation that 'psychopaths' had played an important role in the revolution directly led to the question of how to defend society: 'How can we effectively protect the general public against the dangerous, anti- and asocial psychopathic personalities and mentally inferior? The answer to this question lies equally in criminal law and in practical psychiatry'.[25] The measures proposed by Brennecke mainly consisted in the possibility to detain 'psychopaths' not for juridical or medical reasons, but for the protection of society in specialized institutions under the direction of psychiatrists – an idea that had already been controversially discussed by psychiatrists, lawyers and criminologists in the decade preceding the war.[26] The establishment of specialized institutions for the custody and 'socialization' of 'psychopaths' was also advocated by Kahn.[27] Referring to the psychiatric debates on the reform of criminal law, which had gained new momentum after the end of the war, he highlighted the new importance of psychiatric expertise: 'When during the rearrangement of things our laws undergo the long-planned reform, we will be there to participate and we will not forget what the revolution has told us about psychiatry'.[28] For Kahn, this new claim for social and political expertise also meant that psychiatry's long-lasting process of professionalization was finally complete. Psychiatry, he reminded his fellow doctors in the Munich medical weekly, 'is no longer the poor cousin among the medical disciplines'.[29]

A National Nervous Breakdown

It was not only the actions of a group of anti-social 'psychopaths' that worried many German and Austrian psychiatrists after November 1918. In a 'medical emergency call' (*Ärztlicher Notruf*) published at the end of 1918, Professor Robert Sommer from Gießen warned that the German nation's nervous system itself had suffered a serious shock.[30] With hunger and economic and political crisis driving the German people deeper and deeper into a 'nervous epidemic' (*nervöse Massenkrankheit*), Sommer found the collapse of civilization to be imminent and expected mass suicides, upheaval, overall destruction and, ultimately, the descent into Bolshevism.

In the months to follow, other psychiatrists, among them leading representatives of the discipline such as Emil Kraepelin, Robert Gaupp and, as late as 1923, Karl Bonhoeffer, joined in with Sommer's diagnosis and discussed the current events as symptoms of a collective 'nervous breakdown', 'mass suggestion', collective neurasthenia, psychosis and hysteria.[31] This use of psychiatric categories had been prepared by the rhetorical mobilization of both professional and general public discourses during the war. As early as 1915, Freud had complained about his colleagues' eagerness to diagnose the enemy nations as 'inferior' or 'degenerated'.[32] Three more years of war and a revolution did little to cool down the minds. In 1919, diagnostic terms were ubiquitously used in public political debates throughout all political camps, in the press and in the National Assembly. The 'anti-psychiatric' journal *Irrenrechts-Reform* tried to intervene and clarified: 'There is no such thing as political madness, no war psychosis and no revolutionary psychosis, no legal madness [*Rechtswahnsinn*], and also there is no mass madness'.[33]

When diagnosing contemporary political events, psychiatrists not only proposed easy and seemingly scientific interpretations to an unsettled and disoriented public, but also legitimized a popular discourse with their professional authority. By projecting their medical categories from the individual patient to a collective 'national soul' (*Volksseele*) and by discussing the social and political order in medical terms, psychiatrists extended their expertise on society and the nation as a whole, claiming a formative role in the protection of the nation's collective health and the prevention of future 'hysterical' endemics. Ultimately, this discourse not only added momentum to the bio-political project of eugenics, but also to the establishment and institutionalization of new fields of psychiatric research and practice, namely 'applied psychiatry' and 'mental hygiene'.

The mental state of the collective had played an important role in the examinations of 'psychopaths as revolutionary leaders'. While pathologizing individual protagonists of the revolution, the respective psychiatrists also elaborated on the relationship between the leaders and the crowds, using their diagnoses not only to delegitimize the political claims of the revolutionists, but also to expound their conceptions of the right political and social order. Contrasting the 'psychopathic' leaders of the revolution with the ideal of the 'true leader', they propagated a clear hierarchical order of the state – a model that suited not only the conservatives' wish for the restoration of monarchy or right-wing radicals' hopes for a dictatorial corporative state, but also mirrored a more general preoccupation of Weimar political culture with the figure of the 'leader'.[34]

To answer the question of how 'psychopaths' came to play such important roles during the recent events, Kahn argued, one had to take into account 'the psychology of the two components which, in quiet and in tumultuous times alike, incarnate the lives of the peoples: the psychologies of the leaders and those led, that is, the crowd'.[35] Kahn's thinking about collective psychology was clearly influenced by Gustave Le Bon's popular concept of crowd psychology. A characteristic example of the anti-socialist and elitist positions of late-nineteenth-century French conservatives who saw the political and social order challenged by the emergence of an age of mass politics and the growing influence of the workers' movement, Le Bon's concept could easily be transferred from the French Third Republic to the situation in post-war Germany.

Kahn found the exact opposite of the 'psychopathic' leaders of the revolutionary crowd in the 'true leader', a larger-than-life figure whose psyche was characterized by 'his outstanding creative and critical intelligence, by his unbending, unflinching and pure will and by the total control of all emotions, by the balance of his mind'.[36] Unlike the 'psychopath', whose relation with the crowd is symbiotic, the 'true leader' stands apart from and above the mass of the people, and only because of this total difference is he followed, 'in awe and love, or in hate and fear'.[37] For Kahn, Brennecke, Kraepelin and others the image of the crowd and its 'collective soul' stood in for the mental state of the whole nation. It is here that the anti-democratic and elitist implications of the psychiatrists' discourse on the revolution became most apparent. In the political imagination of many conservative Germans in 1918/19 the incarnation of the 'true leader' was not the former German emperor Wilhelm II, but rather Otto von Bismarck.[38] Hoping for the re-establishment of an authoritarian order, Kraepelin – a representative of the conservative German *Bildungsbürgertum* – projected his ideals of leadership from Bismarck into the future: 'Why should [the German people] not again be able to bring forth a man who can satisfy our longings?'[39]

Some of the psychiatrists who basically agreed with the idea that 'psychopathic personalities' had played an important role in the revolution, declared that this was not a sufficient explanation for the recent events. Robert Gaupp, professor of psychiatry at the University of Tübingen, pointed out that from a medical perspective, it would be unjust to claim that 'the instigation of the masses by radical demagogues was the *only* source of the nameless distress which threatens to swallow Germany'. Instead, he pointed out, the true question was why the 'greatest part of our otherwise so thoughtful and thoroughgoing people has gotten into a state of mind in which it could fall prey to the influence of Russian agents and unscrupulous coffee house writers'.[40] The answer to this question, Gaupp argued, could be found in the collective mental state of both the German army and the people. Hunger, deprivations and suffering both among the fighting troops and on the 'home front' had brought about a mental state well-known from clinical psychiatry: a collective 'neurasthenic' condition caused by fatigue and exhaustion and leading to 'nervous weakness, emotional instability and rootless surrender to the excitement of the moment'. Eventually

> the suffering of the last years, the despair of the lost and costly war, the anger about the years of deception have robbed the quivering nervous psyche of a half-starved people from all interior restraints against the red flood sweeping over it.[41]

After having thus diagnosed the nation, Gaupp suggested a treatment: similar to Sommer, who in his 'medical emergency call' had argued that it was first and foremost the hunger that had driven the German people into 'nervous depression' and 'anarchistic political madness', Gaupp insisted that no recovery was possible without bread and economic and political security. Moreover, he wrote, it was the responsibility of the elites to sacrifice their money and their strength for the benefit of the nation as a whole in order to restore the people's faith 'in its spiritual leaders, [...] the German men and women who by their formation and their education are entitled to win absolute authority and to impart the German culture to the whole of the people'. Without this sacrifice, 'Germany's culture will perish and all will sink into chaos'.[42]

Gaupp's visions of the future were far from limited to a restoration of the Wilhelmine order and its elites. When addressing the medical students of Tübingen on 23 October 1919, he took the medicalization of the political situation to the next level: if political and social problems were caused by individual and collective medical conditions, the only one able to save the nation was the doctor. With a profound sense of mission, Gaupp exclaimed: 'All call for the doctor, the strong-nerved leader [*den starknervigen Führer*] and the saviour of a desperate people'.[43]

In an existential medical and psychological crisis, the doctors had an important role to play because they were the ones who actually knew the people's soul. Propagating a novel and far-reaching expert status for his profession, Gaupp demanded the physicians' 'right to be heard in all public questions'.[44] To save the nation, they had to become the 'educators of the people' and promote its regeneration in many ways: by combatting infant mortality, by propagating marriage and temperance, by opposing abortion, by hygiene education and by calling for a land reform. Moreover, Gaupp not only invoked the importance of medical scientific expertise in all fields of social life but also demanded that physicians acquire charismatic leadership. In a time in which large parts of the population had lost their religious orientation, he saw it the duty of the physicians, and in particular of the psychiatrists, in becoming the spiritual leaders and advisers of the people.[45]

Thomas Mergel has accurately observed that the Weimar Republic's structures of political expectation (*politische Erwartungsstrukturen*) were characterized by a 'constant, sometimes obsessive search for leaders', to the point of 'a messianic search for "Germany's saviour"', who would lead the nation out of degradation and up to new glory' – a desire that was anything but limited to the political right.[46] Yet even against this backdrop, it is striking how Gaupp constructed the doctor-leader as an authoritative public expert, and as an actual alternative to an unfit political leadership. Rejecting both the wartime government and the new democracy, both of which he saw as controlled by a bureaucracy that was ignorant of the people's psychological needs, Gaupp claimed that the 'destiny of our people' finally had to be handed over to those who really understood the people's mind. Against the fragmentation of the nation by interests and parties, he postulated an anti-political vision of a government of medical experts, legitimated by scientific knowledge as well as a deeper understanding of the human condition and a specific ethos of the profession:

> Above all the narrow and antiquated party systems, above all pathetic politics of interest, above all parliamentarian shallowness and vanity, based on a rich knowledge of human nature and a deep love of mankind stands the doctor's way of thinking, which in a daily struggle against poverty and distress and in daily sight of the driving forces of human action teaches how to rightly judge human concerns.[47]

For Gaupp's teacher, Emil Kraepelin, the end of the monarchy and the revolutionary events of the winter of 1918/19 had been a political catastrophe: 'The enormous events that have befallen the German people have deeply shocked its inner life', Kraepelin wrote shortly after the end of the Munich Soviet in an article in the right-wing conservative *Süddeutsche Monatshefte*. Trying to make sense of the recent events, he turned to the diagnostic categories of his discipline and produced one of the most exhaustive psychiatric analyses of the political and social situation in post-war Germany.[48]

Referring to the Paris Commune of 1871 and the Russian Revolution of 1905, Kraepelin saw a historical regularity at work in the current events and applied the categories of clinical psychiatry to the collective level:

> Every persistent and intense pressure on the collective psyche produces stresses which ultimately explode with enormous power and which in their blind rage can no longer be controlled by the forces of reason. In day-to-day psychiatric practice *hysterical disorders* are the counterpart to this behaviour.[49]

To Kraepelin, this analogy was more than just a metaphor: in every mass movement, he pointed out, one could easily find traits which were closely related to hysterical symptoms.[50] As Eric Engstrom has noted, this link between the collective, political events and individual mental disorder played a double role in Kraepelin's argument. By drawing 'the revolution into the clinic', he could not only subject it to a scientific analysis, but bolster his 'psychiatric observations of contemporary events' with his scientific legitimacy as a renowned psychiatrist.[51]

Kraepelin saw more than one reason for the 'hysterical' dimension of the revolution. Apart from the effects of crowd psychology and the leading role of 'psychopathic personalities', he found the participants of the upheaval themselves to be an important factor for the collective hysteria. The revolution had mainly been supported by workers and other members of the lower classes, and in Kraepelin's biologistic worldview class was not a matter of political or economic power relations, but rooted in biological facts.[52] Consequently, he supposed that the revolutionary masses had largely consisted of 'mentally underdeveloped compatriots' (*Volksgenossen*). Unfit to be rational political subjects, they lacked the 'ability for cool calculating consideration, self-control, foreseeing of future events, and the guidance of the will by rational insight'.[53]

The political implications of Kraepelin's polemical article went far beyond the rejection of the revolution and the new political order or a call for the restoration of the pre-war Wilhelmine society. In the rule of the revolutionaries he saw only the last consequence of the belief that all men were equal in their abilities, and only hindered from developing their potentials by external factors such as oppression and exploitation.[54] Kraepelin was convinced that the exact opposite was true and that the stratification of society by and large mirrored the hereditary biological characteristics of its members: on the one hand, he argued, nobility would not have become the ruling class if their ancestors had not had outstanding traits which they could pass on to their descendants. On the other hand, 'it is obvious that the ancestors of those who today belong to the lower social classes by and large did not have any traits that allowed them extraordinary achievements and thus they could not pass down such characteristics'. Nonetheless, this social Darwinist model of a social hierarchy based on a biologic meritocracy was not totally static but allowed for some degree of social mobility: 'We see old and glorious dynasties degenerate and [...] descend into the proletariat', Kraepelin pointed out, and at the same time 'new and vital families emerge without pedigree'.[55] As Eric Engstrom has rightly observed, Kraepelin's social theory was 'conveniently double-edged' and

> reflected the conflicting interests of Kraepelin's own class, the *Bildungsbürgertum*. Confronted with a mass society which ultimately threatened to undermine its own social position, the *Bildungsbürgertum* erected barricades against the supposedly irrational threat from below, while simultaneously ensuring its own asset and hence the selective permeability of the social hierarchy.[56]

'The rule of the people has to become the rule of the best', Kraepelin summarized the consequences of his social theory. Yet, like many other psychiatrists and eugenicists, Kraepelin was convinced that the war had led to a negative selection, robbing the German nation of 'the men most gifted and most willing to sacrifice themselves' while sparing 'the unable and self-serving'.[57] As he did not want to content himself with the best to emerge by chance or nature, he advocated an active intervention and a far-reaching programme for the recovery of the nation. In order to avert degeneration, Kraepelin proposed a number of measures, most of which had already been part of social hygienists' and Kraepelin's own agenda before the war: early marriage, fertility, the fight against alcohol, syphilis and the distresses of urban life. What was necessary now, he wrote, was 'by all means, to breed outstanding personalities who in the arduous days to come, may guide our fortunes'.[58] But at the same time, Kraepelin also stated that the 'good parts of our *Volk* should not be ruined by the inferior ones' and that the 'inferiors' should not be a burden to the national collective. Here, Kraepelin's socio-political ideas already show the outlines of a 'negative' approach to eugenics which, following another radicalization after the World Economy Crisis, would ultimately lead to the forced sterilizations and the 'euthanasia' programme of Nazi psychiatry.[59]

Yet, the most vociferous propagandist of psychiatry's claim to political and social expertise was Erwin Stransky, professor of psychiatry at the University of Vienna. Shortly before the end of the war, Stransky had already published his manifesto for a new way of psychiatric research and practice which he had labelled 'applied psychiatry' (*Angewandte Psychiatrie*). Even more explicitly than Gaupp or Kraepelin, he found the psychiatrist to be the ultimate social and political expert:

> There is no other human being, no other physician, no one, whose work would allow him such deep insights into the deepest psychic matters of life, of individual men, of groups of men and even of the peoples ... than the psychiatrist!

Yet, Stransky asserted, only too few psychiatrists were aware of the potentials and responsibilities of their profession and most of them remained stuck in the unworldly isolation of the asylum and the laboratory.[60] Sharply criticising his colleagues for their readiness to compromise and their lacking self-confidence, he called for a 'healthy imperialism of the doctors' in the service of the protection of society and of racial hygiene.[61] As a first objective in psychiatrists' campaign for 'power politics', Stransky propagated the conquest of the legal system. Step–by–step, psychiatrists had to expand their current status as expert witnesses, up until the 'dethronement of law' (*Jurismus*): 'Historia docet! After Pippin followed Charlemagne and the sons of today's consulting experts will be tomorrow's leaders and judges of mankind'.[62] Going beyond mere rhetoric, Stransky – a dedicated propagandist of ethnic Pan-German nationalism in Austria – merged the language of national power politics with the professional policies of his discipline, calling for an 'imperialism of the doctors', professional 'power politics' (*Machtpolitik*) and a '*großärztliche* propaganda' (greater medical propaganda).[63]

The expansion of psychiatry's field of activity as envisaged in the agenda of 'applied psychiatry' was not limited to the conquest of institutions and the strongly increasing presence of psychiatric expertise in all areas of social and political life. In order to become what Stransky called (with a hardly translatable German concept – a 'five-star' general expert) '*Generaloberstsachverständiger*' for all forms and ways of life of the individual and the collective, the psychiatrist had to open up new fields of research.[64] In particular, the topics of ethnography and the social sciences had to be re-examined from a psychiatric perspective in order for the psychiatrist to become the 'teacher and guide of future statesmen and diplomats'.[65] History was to play a particularly important role for the psycho-political expertise offered by 'applied psychiatry': whereas the historians, ethnologists (*Kulturforscher*), economists and politicians lacked the psychological knowledge to learn anything valuable from history, Stransky found the psychiatrist to be the one who could understand history and draw the right conclusions for the future.[66]

What had already been a far-reaching agenda for the renewal of the psychiatric profession and for the extension of its field of activity seemed even more urgent after the military defeat and the ensuing upheaval. To Stransky, the immediate post-war period had made even clearer how important the understanding of 'practical psychology' (*Seelenkunde*) by both the people and its leaders would have been to avoid this 'gruesome catastrophe'. Lecturing at the meeting of German psychiatrists in Hamburg in May 1920, he renewed his call for 'applied psychiatry', which now had to be placed in the service of the 'mental reconstruction of the German people'.[67] 'Applied psychiatry' was now redrafted as an expansive programme for the bio-political and psycho-political reform and re-education of the common people and the elites. Many of the demands that Stransky presented in a characteristically overheated rhetoric differed neither in kind nor in degree from the socio-medical interventions propagated by Gaupp, Kraepelin and others: education of the people in *völkisch* (ethno-racialist) virtues, positive eugenics, temperance and the fight against syphilis. Yet, more than other doctors, Stransky not only advocated an expansion of psychiatrists' socio-political expertise, but also postulated that psychiatry had to change its professional profile in order to claim this authority.

It was Stransky's far-reaching plan for the expansion of the discipline's psycho-political expertise that provoked one of the few critical public statements by a fellow psychiatrist. In a 1921 article, Arthur Kronfeld used Stransky's approach as an occasion for a broadside against the more general tendency to extend the reach of psychiatric diagnoses into social and political matters, describing 'applied psychiatry' as one of the greatest threats to 'the objective and logical integrity of our discipline'.[68] Although he insisted that he was motivated only by a concern for the objectivity of science, his rejection of 'applied psychiatry' was also due to political reasons: Kronfeld was a member of the Social Democratic Party and, in late 1918, had been a delegate in the Freiburg Soviet, a political alignment that strongly differed from the prevalent right-wing nationalism of German psychiatrists.[69]

Outlook: Psychiatric Expertise in the Interwar Period

The persuasiveness of psychiatrists' visions of a national nervous breakdown and rampaging 'psychopaths' was closely linked to the specifics of the post-war situation. With the gradual economic and political consolidation of the Weimar republic and the Austrian First Republic during the 1920s, this kind of explicit socio-political diagnosis largely disappeared from the major psychiatric and neurological journals, while conservative psychiatrists reluctantly made their peace with the new state of things. Nonetheless, the general idea of translating the concepts of psychopathology into tools for the analysis of society remained very much alive. Apart from the activities of Erwin Stransky and the Association for Applied Psychopathology and Psychology in Vienna, examples also include Karl Birnbaum's layout of a 'psychopathology of culture' (*Kulturpsychopathologie*) published in 1924, or the works of Arthur Kronfeld, who – despite his severe criticism of Stranksy's approach – ventured into the field of 'sociological psychopathology' and 'psychopathological sociology' in 1923.[70] However, the most influential actualization of Le Bon's ideas on mass psychology came from the young discipline of psychoanalysis. Sigmund Freud's seminal *Massenpsychologie und Ich-Analyse* dates from 1921 and can be read in the wider context of psychiatrists' increased interest in socio-political matters in the early interwar years.[71]

Beyond the boundaries of the medical and 'psy'-disciplines, the idea of a malady of the collective body became one of the essential topoi of right-wing conservative discourses in the interwar period. Conservatives of every shade commonly evoked the image of national illness and national regeneration in order to advocate their political agendas.[72] This discourse often had a distinctly psychiatric dimension: the frequent use of concepts like the 'national soul' or collective nervousness shows how psychiatric knowledge had been adopted by a wider public. Although the use of such concepts can be traced back at least to the first third of the nineteenth century, the events of the post-war period endowed this discourse with both urgency and plausibility.[73] When Hermann Oppenheim and Emil Kraepelin published their diagnoses of the revolution in high-circulation media like the *Berliner Tageblatt* and the *Süddeutsche Monatshefte* respectively, they clearly had this kind of dissemination into a broader educated public in mind.[74] Yet, what direct impact their ideas had and if their authors could capitalize on them in terms of social and scientific prestige remains difficult to assess.

As a discipline, psychiatry could successfully consolidate both its standing as a scientific discipline and its role as an interpretative authority in social and political affairs during the interwar period. Apart from the creation of new university departments and the expansion of existing ones, the incorporation of the German Research Institute for Psychiatric Research (*Deutsche Forschungsanstalt für Psychiatrie*, DFA) in Munich into the Kaiser Wilhelm Society in 1924 reveals the increasing scientific and socio-political relevance of psychiatry.[75] Founded in Munich in 1917, the DFA quickly became one of the most important institutions in German psychiatry even before its integration into the major umbrella organization for scientific research in Germany. Its creation was, more than anything else, the result of the organizational efforts of Emil Kraepelin, who had already begun campaigning for a psychiatric research institute in 1912. Kraepelin's lobbying for a psychiatric research institute was closely linked to his social and political ideas. When presenting plans for the future institute in 1915, the main reason he gave for its creation was the necessity of fundamental research in psychiatry for the fight against the 'devastations that mental illness causes to our national body'.[76] Even more than about the mentally ill he was concerned about the many 'slightly abnormal people, who we describe as "nervous", eccentrics, psychopaths, or as feeble-minded, inferiors, degenerates, and enemies of society'.[77] However, he argued, the 'weapons' against these dangers which threatened the very existence of the nation and society could not be developed in the messy, everyday practice of insufficiently equipped psychiatric clinics and asylums.

Kraepelin's plan for a research institute moved considerably closer to its realization in 1916, when a large donation by the Jewish-American philanthropist James Loeb provided a financial basis. In April 1918, practical research activities by the DFA commenced, shortly before the end of the First World War and the revolution in Munich. Both Emil Kraepelin and Eugen Kahn – members of the institute's staff – were among the most aggressive psychiatric commentators on the 1918/19 revolution. During the interwar period, the institute became a national and international centre of psychiatric and neurological research. Kraepelin's emphasis on the role of psychiatry in the process of national regeneration was reflected in the DFA's organizational structure, which, since the institute's foundation, included a department for genealogy and demography led by Ernst Rüdin, one of the most important representatives of racial hygiene and eugenics in Germany.[78] Under the direction of Rüdin, who was appointed head of the DFA in 1931, research in heredity, eugenics, and genetics, as well as in criminal biology and 'psychopathy', became increasingly important for the whole institute.[79] After 1933, he was one of the most important psychiatric experts in the Third Reich, and played an important role in the scientific legitimization of the National Socialists' medical policies, both domestically and abroad.[80]

The interwar period opened up new possibilities for expert activity, and psychiatrists were able to occupy important positions. In particular, the expansion – and bureaucratization – of social welfare created a new market for scientific expertise from different disciplines, both in Weimar Germany and in Austria.[81] The question of which pensions mentally injured veterans should be entitled to had already been one of the key topics in the controversial debates on the so-called 'war neuroses' before 1918. In Germany, the passing of the pension law (*Reichsversorgungsgesetz*) in 1920 finally promised pensions in the case of mental disorder due to war experiences or work accidents. But as crucial passages of the law were relatively open to interpretation, its implementation required the participation of psychiatric experts on all levels, from testimonies in individual cases to high-level policy advice. With health officials in need of expertise and psychiatrists eager to extend their sociopolitical influence, the welfare system of the early Weimar Republic may well be described as a situation in which science and politics functioned as 'resources for each other'.[82] As Stephanie Neuner has recently shown, the pension question mobilized a highly active and stable network of health officials and psychiatric experts, in which a small and exclusive circle of conservative psychiatrists was able to exert some influence on national health and welfare policies in Germany.[83]

Another field on which psychiatrists could defend, and in some regards even extend, their expert status was the judicial system. Beyond the continuous importance of forensic expert testimonies in the court room, the post-war period saw an increasing institutionalization of psychiatric expert knowledge in the penal system. The debates led by psychiatrists during and after the war had a strong influence on the theory and practice of criminology in the interwar period. Richard F. Wetzell has shown that the increasing importance of 'criminal biology' in Weimar Germany was closely connected to psychiatry's expansion beyond the clinic and into society, as well as to psychiatrists' increasing concern 'with the welfare and protection of society as a whole rather than the individual patient'.[84]

More than anything else, the concepts of 'mental inferiority' and 'psychopathy' were a crucial factor in the gradual 'medicalization of penal law', constructing a criminal, moral and political menace to bourgeois society that only trained specialists could safely identify and assess.[85] Notably, the allegation that 'psychopaths' had been the protagonists of the revolution was also adopted by political decision makers in the post-war period. In September 1920, the Prussian minister of welfare, Adam Stegerwald, prompted the establishment of specialized counselling offices for 'psychopaths'. Although these offices were clearly not the institutions for the custody of 'psychopaths' that Eugen Kahn, Hans Brennecke and others had envisioned, Stegerwald basically used the same arguments to back his initiative: the recent upheaval, he wrote, had most clearly shown that 'juvenile psychopaths are to be found in the frontline of politically extreme movements'. Yet, although the 'psychopaths' were thought to be a threat to the whole nation, Stegerwald was convinced that forced medical treatments would not be successful. In contrast, counselling offices, which had to be strictly separated from the asylums, would offer the possibility to more effectively reach and treat 'psychopathic' individuals with their own consent.[86]

Counselling offices for 'psychopaths' were but one outcome of a more general tendency towards the prophylaxis and prevention of mental illness. Against the background of psychiatry's notorious inability to heal its patients, psychiatrists discussed a wide range of different approaches for the prevention of mental illness and the preservation of both individual and collective mental health. As I have shown in the previous sections, different forms of socio-medical interventions had already been an integral part of psychiatrists' psycho-political diagnoses of the immediate post-war period, where the threat of mass-hysterical endemics and the need for national regeneration could serve as a legitimization for the expansion of psychiatry's field of activity. Although the debate lost some of its alarmist edge, it continued during the interwar years and led to the emergence of a movement for *psychische Hygiene* or *Psychohygiene* ('mental hygiene') in the German-speaking countries. Together with parallel and related projects in many European and non-European countries, it was part of an international movement for 'mental hygiene', founded in the United States before the war.[87] An important step in the institutionalization of this loosely defined concept was the creation of a German Society for Mental Hygiene (*Verband für psychische Hygiene*) in 1925 through the initiative of Robert Sommer, who also became its first president. The first conference of the society, held in Hamburg in 1928 not only attracted leading psychiatrists but also state officials and representatives of the welfare authorities and the police.[88]

Many of mental hygiene's approaches to collective mental health prophylaxis, such as outpatient care, counselling and recreation, were soon gradually pushed aside by calls for more resolute socio-medical interventions. As the example of the Society for Mental Hygiene shows, the concept of mental hygiene itself was increasingly reduced to eugenics. Since the establishment of the society, its concept of mental hygiene and the prophylaxis of mental illness had explicitly comprised eugenics. These concepts had already been prevalent among German psychiatrists before and during the war and had gained additional momentum against the backdrop of the war and the post-war crisis. As Paul Weindling has argued: 'Virtually any aspect of eugenic thought and practice – from "euthanasia" of the unfit and compulsory sterilization to positive welfare – was developed during the turmoil of the crucial years between 1918 and 1924'.[89] Yet, eugenic concepts only came to dominate the psychiatric discourse both inside and outside of the Society for Mental Hygiene in the late 1920s, when the impact of the World Economy Crisis increased the economic pressure on the welfare system and brought an end to many reform-oriented projects in psychiatry.[90] In the early 1930s, eugenic thinking increasingly displaced alternative approaches to psychiatric prophylaxis. After the Nazis' rise to power, the society's understanding of mental hygiene became more and more indistinguishable from racial hygiene and eugenics, since Ernst Rüdin was appointed president of the society.

Conclusion

What can the example of psychiatrists' psycho-political diagnoses in the aftermath of the First World War tell us about the history of scientific expertise in the first half of the twentieth century? First, it certainly provides us with one of the most striking examples of the connection of scientific expertise with political commentary, and of how inherently political psychiatric categories of normalcy and deviance could become explicitly politicized. Second, it clearly shows the ambivalent situation faced by scientific experts at the end of the war: in the case of psychiatry, the war had offered considerable chances for the improvement of the discipline's standing. The epidemic of so called 'war neuroses' had made psychiatry an essential part of the war effort, and their apparently successful treatment promised the end of psychiatry's 'therapeutic nihilism'. Defeat and revolution threatened to unravel these wartime gains: the conflicts created by the often brutal treatment of 'hysterical' soldiers erupted both in the clinics and in political debates, while at the same time psychiatrists anxiously observed the dissolution of the old social and political order. Nonetheless, the post-war situation presented considerable chances for mental health experts: the extension of the welfare state offered new possibilities to implement socio-medical interventions, which were further legitimized by the prevalent rhetoric of crisis and 'national reconstruction'.

In this context, psychiatrists' psycho-political diagnoses served a double function. On the one hand, the spectre of the revolutionary 'psychopath' and a national 'nervous breakdown' gave utterance to the fears of an educated middle class, while at the same time delegitimizing its political adversaries and their claims. On the other hand however, behind the bleak pessimism of these diagnoses, psychiatrists tried to seize the opportunities that the situation offered and called for large-scale public health programmes under their own leadership. Based on the claim that psychiatry had a privileged insight into all human affairs, they presented themselves as the only ones truly able to analyse and understand the current situation – and to prevent its future repetition. To some extent this strategy was successful, and psychiatric proposals of the immediate post-war period as counselling offices for 'psychopaths' were taken up by state officials. However, the 'active treatment' of wartime psychiatry had not led to the hoped-for therapeutic breakthrough. During the interwar period, the promise of prophylaxis was the strategy of choice to maintain psychiatry's position in the contested field of public health. Although other approaches were also discussed, the eugenic paradigm proved to be the most successful, securing considerable funding for psychiatric research institutions long before it became part of the official 'racial hygiene' policy of the Nazi state.[91]

6 CONTESTED MODERNITY: A. G. DOIARENKO AND THE TRAJECTORIES OF AGRICULTURAL EXPERTISE IN LATE IMPERIAL AND SOVIET RUSSIA

Katja Bruisch

When, in 1926, A. G. Doiarenko celebrated his twenty-fifth jubilee as an agricultural scientist, he ranked among the most prominent agricultural specialists in Soviet Russia.[1] An authority in the field of agricultural soil science, he held a professorship at the K. A. Timiriazev Moscow Agricultural Academy and headed the academy's experimental station. Numerous congratulatory telegrams reflected the considerable respect that Doiarenko enjoyed in the scientific community. His colleagues praised the professor as a leading scientist in his field.[2] A group of his former students stressed Doiarenko's efforts in bridging the gap between science and agriculture, lauding his close ties with the rural population:

> Science never removed you from real life. All these 25 years, you have stood, your 'face to the village', considering this to be natural in an agricultural country ... United in your character are the scholar, the educator and the public figure. You have served the aim of increasing crop yields while also endeavouring to improve the lives of the peasants.[3]

Doiarenko was also honoured in public. In a newspaper article, a colleague called him an 'indispensable member' of the scientific staff of the People's Commissariat for Agriculture (*Narodnyi komissariat zemledeliia* – Narkomzem), the highest-ranking agency for agrarian policy in the country. The author of the article also mentioned that there was a discussion in the Narkomzem about giving special recognition to Doiarenko for his merits.[4]

The spectrum of this praise reflected the wide range of Doiarenko's professional engagement and the different social stages on which he acted as an expert. After finishing his agricultural studies, he had not only devoted himself to scientific research but had also supported various initiatives for popularizing scientific knowledge. The desire for furthering dialogue between scientists and the rural population was a thread throughout his career. After completing his studies at the Moscow Agricultural Institute, Doiarenko assisted in preparing lectures on agriculture for peasant farmers.[5] In the 1920s, he achieved a certain fame for his 'peasant lectures' (*krest'ianskie besedy*): on ten Sundays during the year, he invited peasant farmers from the Moscow environs to the Moscow Agricultural Academy in order to teach them about fertilizers, new methods of cultivation or how to best use various tools and instruments.[6] At the same time, Doiarenko was involved in the agricultural policy of the Soviet state. In the 1920s, he advised the People's Commissariat for Agriculture on questions of agricultural experimentation and was active as a member of the Department of Agricultural Planning within the Narkomzem.[7]

From the perspective of a Russian scholar of the time, these various social roles did not exclude each other. With the rise of the so-called 'agrarian question' in the late nineteenth century, when the modernization of agriculture came to be seen as one of Russia's major challenges, people trained in agricultural sciences or related disciplines often combined an academic career with public engagement and political consultancy. In this chapter, Doiarenko's professional biography will serve as a lens through which to explore the mechanisms underlying the social construction of expertise within such different political regimes as the Tsarist and the Soviet. Special attention is given to the questions of to what extent historical breaks affected the way in which scholars performed their expertise, how they perceived themselves and how effective they were in shaping the dominant discourse about the countryside and the agrarian policy of the state. In examining the mutual relations between the scientific, the public and the political spheres, the present analysis reflects on the link between expertise and the various projects of modernity in late Imperial and Soviet Russia.

Knowledge Production and the Quest for Social Transformation

Doiarenko's commitment to the peasantry was part of a longer tradition going back to the second half of the nineteenth century. At its threshold stood the 'discovery of the peasantry' in the context of the abolition of serfdom by Tsar Alexander II in 1861. With the strengthening of the village communes and the formal establishment of a peasant jurisdiction separate from the state organs of justice, the peasantry became visible as a distinguishable social group.[8] From now on, all kinds of social and national aspirations and blueprints for change would be projected on the rural population. Historians, lawyers and ethnographers seized upon the peasantry as their special object of study. Writers and artists made the 'Russian peasant' a preferred focus of their own creativity, while the tsarist state officials shifted it into the very center of their political rhetoric. Activists in the social-revolutionary movement (*narodnichestvo*), who saw in the village community a prototype of socialism, revered the peasants as the potential agents of revolutionary upheaval. Conservatives stylized them as the solid bedrock of autocracy and orthodoxy. Their ideological and political differences notwithstanding, the non-peasant elites agreed on one decisive point: they believed that the peasantry constituted a specific social group upon which the ultimate fate of the Russian Empire depended.[9]

The distinguishing of the legal norms and customs of the village population from those of other social groups went alongside the idea that there existed a specific peasant mode of agriculture. While agricultural scientists had up until that juncture concentrated their attention on the estates of the landed gentry, they now became interested in the farming methods of the 'people'. Decisive for this conceptual shift were the statistical surveys carried out by thousands of statisticians employed within the organs of local self-government (*zemstvos*) set up in 1864.[10] Originally invented to quantify the taxable income of the rural population, *zemstvo* statistics became the empirical basis of an economic theory of peasant agriculture, given its most elaborate expression in the work of the agrarian economist A. V. Chaianov in the early twentieth century.[11] As a result, the natural-scientific paradigm which so far had dominated academic agrarian discourse was replaced by a sociological approach towards the study of the rural economy. While in 1837 the professor for Mineralogy, Physics and Agriculture M. G. Pavlov had maintained that 'agriculture as a science' meant 'the application of the natural sciences in an aim to raise plants and animals, which are of common use',[12] the famous agronomist Aleksei F. Fortunatov, a former *zemstvo* statistician, taught his students in 1903 that natural science and social science were 'two equally fundamental foundations of agronomy'.[13]

Underlying the establishment of agronomy as a social science was an agenda devoted to the transformation of the social hierarchies of the Russian Empire. Scholars of the countryside no longer attributed agricultural success solely to the correct application of the natural sciences and sophisticated methods of accounting, but rather considered the social conditions of farming a major factor in agricultural development. As a result, these social conditions themselves became an issue within intellectual debates on the rural economy. The quest for social transformation was inherent to *zemstvo* statistics from the beginning. Influenced by the ideas of the *narodnichestvo*, the statisticians expressed a strong commitment towards the rural population. Unlike the initial followers of the populist movement, who had sought to mobilize the peasantry for the sake of revolution, the *zemstvo* statisticians wished to improve the living conditions in the countryside by means of so-called 'small deeds' (*malye dela*). These were seen as a first step toward the integration of the peasantry into the larger project of social change.[14] In the context of the nationwide peasant unrest in 1905/06, agrarian scientists and political economists discussed social conditions of agriculture from a macro-sociological perspective, arguing that agricultural growth presupposed a redistribution of agricultural land to the benefit of the peasant population.[15] Even if the question of land distribution receded into the background in the period before World War One, and projects favouring the modernization of agriculture gave stronger emphasis to the spread of knowledge and technology, the transformation of the empire's social hierarchies remained a paramount concern for many scholars of agricultural sciences and related disciplines. In the tradition of the *narodnichestvo*, they expressed a strong sense of responsibility toward the 'people', thus presenting themselves as representatives of the common good.[16] As A. F. Fortunatov put it, the task of the agronomist was not limited to supporting the rural population in agricultural questions. Rather, he should also promote the 'intellectual and civil development' of the peasants (*umstvennoe i grazhdanstvennoe razvitie*), bringing them 'bread' (*khleb*), 'enlightenment' (*svet*) and 'freedom' (*svoboda*).[17] Fortunatov's view is symptomatic of the self-perception of many of his colleagues at the time. Considering their task to liberate the peasants from their marginality by means of education and enlightenment, they presented themselves as the champions of a new social order in which the peasants would participate as equals.

In the months following the February Revolution in 1917, the future of the Russian countryside was one of the most debated political issues. Under these new conditions many scholars of the countryside claimed to act as advocates of a marginalized peasantry whose interests they would represent at the level of state policy. Referring to the economic viability of peasant households, they enthusiastically endorsed the idea of making the 'toiling peasantry'[18] the bedrock foundation of Russian agriculture. Doiarenko, who became one of numerous agricultural advisers to the Provisional Government, stressed that the maximizing of agricultural production was not the only objective of agrarian reform:

> If the task we had to accomplish was one of an exclusively organizational nature or involved solely technical aspects of production, we could also arrive at organizational conceptions totally alien to the principle of 'all land to the toiling people', a principle oriented to fair and equitable distribution – and not to maximizing of production.

Accordingly, the country's rural order had to be based on the 'ideological principles of the broad masses of the people'.[19] Doiarenko's words express the extent to which for contemporary scholars it was a matter to act as academicians and as proponents of certain political interests at the same time. Drawing a connection between the social order and the development of agriculture, the protagonists of academic agrarian discourse rendered science the point of departure for criticism of the marginal position of the peasant. In early twentieth century Russia, science and the quest for social transformation were intrinsically linked to each other.

Scholars into Experts

The scholars of the peasantry became experts due to the rising political and public interest in the countryside at the turn of the twentieth century. As a result of the 1891/92 famine the peasants were increasingly regarded as a population group that had to be familiarized with the achievements of modern civilization.[20] In this context, agriculture became a major policy concern. After the reform of the Ministry of Agriculture completed in 1894, the tsarist government embarked on an interventionist agrarian policy[21] culminating in the so-called Stolypin reforms in the early twentieth century.[22] The 'agrarian question' was also a focus of animated discussion beyond the circles of government. In Moscow and St. Petersburg, as well as in the provinces, agricultural associations recorded a rapid rise in membership, while the number and circulation figures of magazines and popular-scientific publications dealing with agriculture soared.[23] For those active in the agricultural sciences and their neighbouring disciplines, these developments meant a growth in professional opportunities and social authority. The famine occurred at a time when the belief that the world could be shaped by knowledge and reason had become a dominant theme in elite discourse.[24] In the face of the widespread conviction that agricultural growth could be generated by conscious intervention, scholars were increasingly demanded as experts; 'persons who, based on their routine contact with specific topics, are assumed to have accumulated experience in contexts relevant for taking action, and thus enjoy both trust and social respect'.[25] A clear indicator for this was the rapidly expanding labour market for agricultural specialists. In the decade before World War One, thousands of agronomists, statisticians, surveyors, veterinarians and economists found employment in the local branches of the Ministry of Agriculture, in *zemstvos* and the central organs of the expanding cooperative movement.[26]

Doiarenko's career trajectory is a perfect illustration of how rural Russia shifted into the centre of public attention and how knowledge related to agriculture became a source of upward social mobility. In his youth, Doiarenko probably could never have imagined that for his entire life he would be dealing, of all things, with questions of agriculture. The son of a housemaid from the Ukraine, he had initially dreamed of a career as an engineer. After he failed in his bid for admission to the St. Petersburg Transport Institute, he studied natural sciences and law at the university in St. Petersburg. Subsequently, he also completed an advanced course in composition under the famous composer, Nikolai A. Rimskii-Korsakov. His decision to become involved with problems of agriculture came in the late 1890s, when the modernization of rural Russia turned into a major public concern. In 1898, Doiarenko began his studies at the Moscow Agricultural Institute, the institutional successor of the famous Petrovka Academy for Agriculture and Forestry founded in 1865. In his brief historical account of the institute, published in 1915, Doiarenko mentioned that in the late 1890s the famine had significantly contributed to the 'awakening of the society'.[27] Apparently, the public commitment to the countryside which developed in the immediate aftermath of the catastrophe was a point of reference also for his professional self-understanding as an agronomist.

Scholars like Doiarenko were quite successful in legitimizing their new status as experts. In the years prior to World War One, they successfully sought to institutionalize their interpretation of the agrarian question and to communicate it to a growing public audience. This became possible thanks to the expansion of the agricultural educational system promoted by the government and the increasing importance of private institutions which served to compensate for the shortage of agricultural study programmes.[28] Targeting individuals whose educational qualifications did not permit them to study at state universities and colleges, or who were excluded from such study for legal reasons, these private institutions became places where scientific agrarian discourse was extended beyond the milieu of its initial representatives.[29] Doiarenko's professional career was typical in this sense as well. Along with his appointment at the Moscow Agricultural Institute, he was a lecturer at the Golitsyn Agricultural Courses for Women in Moscow, founded in 1908, whose student numbers soared from thirty in 1908 to more than 900 by 1915.[30] He simultaneously taught at the Shaniavskii Moscow City People's University, where in the years before World War One, all the famous protagonists of the academic discourse on agriculture gathered.[31] Here too the scholars found rapidly rising student numbers, rocketing from 1,106 in the academic year 1909–10 to 6,442 in 1914–15.[32] Periodicals were another important means for expanding the scope of scientific agrarian discourse. For example, Doiarenko edited the 'Agricultural Messenger' (*Vestnik sel'skogo khoziaistva*). Run by the Moscow Agricultural Society, this journal served as one of Russia's most significant forums for leading scholars, as well as practical agronomists in the provinces, until the Moscow Agricultural Society was closed down by the Soviet government in the late 1920s.

As a result of the institutionalization of the 'agrarian question', scholars in disciplines related to agriculture developed a collective identity. Like other professional groups in late imperial Russia, they forged a supra-regional space of communication in which they debated their ideas and developed strategies for joint public action.[33] The Society for the Mutual Assistance of Russian Agronomists, established on the initiative of agricultural scientist I. A. Stebut, organized training programmes for agricultural specialists. The annual conferences of the society, attended at times by more than 700 participants, contributed to the continuation of professional contacts on a regular basis. Doiarenko, who was an active member of the society, edited a periodically published list of active agronomists in the country.[34] The Moscow Society for the Dissemination of Agricultural

Knowledge Among the People, mainly involved with the organization of courses for peasant farmers, helped to obtain the papers needed for official accreditation as an agronomist. It thus functioned as a professional body through which agricultural specialists could lobby the state authorities.[35] Regular congresses of agronomists and activists from the cooperative movement served the representatives of scientific agrarian discourse as a basis for the articulation of their interests, both to a broader public and to the agencies of the central state.[36]

The upgrading of expert knowledge on agriculture and the village economy had a decisive impact on the mode in which scholars of related fields described themselves. Faith in science and optimism about the future merged in an elitist sense of mission. One of its most vivid testimonials was a poem by A. F. Fortunatov, which the members of the Moscow Agricultural Institute made into their hymn, furnished with music by A. G. Doiarenko: 'There is a faith in us and this faith is our strength, that the future belongs to us'.[37] Trust in the power of scientific knowledge also changed the way in which the scholars positioned themselves vis-à-vis the rural agrarian population. In the decades preceding World War One, the philosophical conception of the peasantry lost its power. Instead, a civilizing mission gained importance, which fused the ethos of Russian populism with a modern faith in the transforming power of science. As a result, the self-image of the intelligentsia as liberators was replaced by the idealization of the modernizing expert.[38] In the decade before the First World War, the figure of the agrarian expert embodied the widespread dream of transforming agricultural practice through the agency of reason. In 1907, students from the Moscow Agricultural Institute created a student circle, designed to further exchange between students, professional agronomists and professors. A short report on the activity of the study group contained a succinct definition of the professional roles of the agronomist. He was a 'propagandist, a popularizer, a pedagogue'.[39] In practice, this self-image became manifest in numerous initiatives by agronomists to spread and popularize scientific agricultural knowledge through travelling exhibitions or lecture series for peasants, such as those carried out by the Moscow Society for the Dissemination of Agricultural Knowledge among the People, whose executive board Doiarenko belonged to.[40]

Despite the rhetorical imaging of the agronomist as a trailblazer of agricultural modernization, the distinction between experts and laypersons established in the intellectual discourse about the Russian village was not unambiguous. The conviction that solving the agrarian question required educating the peasantry did not automatically imply that the intellectual elites saw the rural population as some sort of silent mass. They rather stressed the peasants' legitimate interests and relevant experiences. As the economist A. V. Chaianov put it in 1915: 'Scientific laws cannot replace the mind of the farmer … they rather stir it up to work intensively. Only with the farmer's mind will scientifically organized agriculture bear fruit'.[41] Such arguments remained in place also after the Bolsheviks had taken power. When in 1925 Doiarenko published a handbook for potential organizers of agricultural lectures meant for peasants, he took a stance quite similar to that of Chaianov ten years before:

> 1. The peasant audience possesses a huge amount of knowledge and practical experience; in many cases, this exceeds the lecturer's knowledge.

> 2. On the other hand, the lecturer has knowledge that the peasants are in need of. He also has the ability to explain and clarify difficult agricultural problems.

> 3. The task of our peasant lectures is to bring together these two different sources of knowledge, practical experience and theoretical thinking, to discuss every measure being taken by the peasants and to define the correct ways for agriculture.[42]

These assertions prove that, contrary to what has been argued, in the discourse of the elites the peasants were more than 'objects of action rather than participants in their own transformation'.[43] In Russia, as in Germany or Italy at the same time, prominent agricultural experts showed a high degree of respect towards local knowledge and practical experience.[44] In imagining the outcome of any agricultural modernization programme to be dependent on an exchange between science and practical agriculture, many scholars treated the peasants as actors whose involvement was as decisive for agricultural success as the experts' initiatives.

Knowledge as a Political and Administrative Resource

The reputation and influence of the experts was inseparably linked to the political turning points in early twentieth-century Russia. The First World War, which in the Russian Empire, like in other warring states, led to an expansion of state intervention in the national economy, transformed the social role of agricultural specialists. In view of the crisis in food supply, the government recognized the administrative and political value of knowledge related to agriculture. Within the so-called 'parastatal complex', a 'dense network of professional and civic organizations that became closely intertwined with the state',[45] numerous scholars of agriculture and the rural economy participated in the attempts to regulate the Russian economy in the interest of conducting the war.[46] Doiarenko, for example, was a member of the Economic Section of the All-Russian Union of Towns, which dealt with the problem of rising food prices and developed plans for the supplying of basic provisions to the army.[47] For the scientists, who saw in their new activity a possibility for realizing an economy organized on the basis of science and reason,[48] the new responsibilities served as a source of additional professional legitimacy: if before they had acted primarily in the forums of a social public sphere critical of the government as well as the empire's social and political order, they now became officially recognized experts whose knowledge was regarded as a decisive factor for the successful continuation of the war.

After the collapse of the tsarist government in early 1917, the hour of the experts appeared to have arrived: the Provisional Government, which inherited the supply crisis of the former regime and at the same time committed itself to implementing a comprehensive agrarian reform, included numerous members of the professional and academic elites within its structures. Doiarenko became a member of the council of the League for Agrarian Reforms (*Liga agrarnykh reform*), an advisory committee for agrarian policy that leading economists, statisticians and agrarian scientists had established shortly after the abdication of the tsar.[49] In addition, he belonged to the Main Land Committee (*Glavnyi zemel'nyi komitet*), which was instructed to draft the guidelines for land reform, and in which economists, statisticians and agrarian scientists occupied leading positions.[50] The growing number of scholars and scientists within the government and the state administration symbolized the optimistic faith in the possibility of realizing a policy based on objective facts. At the same time, the inclusion of the experts in the state policy served as a means of underscoring the idea that the Provisional Government was making an honest attempt to take the interests of the broader population into account. The experts themselves were enthusiastic about this new configuration of science and politics. As the economist A. V. Chaianov put it: 'The agrarian question has now shifted from the realm of abstract ideas and principles into that of practical work in economic organizing'.[51] Economic and socio-political visions, it seemed, would now be fruitfully fused.

The October Revolution led to the successive nationalization of agricultural expertise. Continuing with the expansion of the interventionist state that had begun during the First World War, the Bolsheviks promoted the integration of professional elites into the state administration.[52] At the same time, they began including public spaces into a 'Soviet public sphere' controlled by the party and the state.[53] In the course of these developments, the agricultural experts found themselves confronted with changing institutional conditions for collective public action. Their traditional fields of activity disappeared as local self-government was abolished. Moreover, the central government pressed forward with the nationalization of the cooperatives and the incorporation of public educational institutions into the structures of the state education system. Numerous economists, statisticians and agricultural scientists who had no ties with the Bolshevik party at all, now became part of the Soviet state apparatus, especially the People's Commissariat for Agriculture.[54] Some of them had come to an arrangement with the Bolsheviks before the civil war had even finished. In August 1918, Doiarenko, at the time professor at the Moscow Agricultural Academy, A. P. Levitskii, head of the experimental station for the Moscow area, and V. P. Kochetkov, long-time member of the Society for the Dissemination of

Agricultural Knowledge among the People, began advising the Narkomzem on questions of agricultural experimentation.[55]

Although as typical representatives of the professional intelligentsia, many agricultural experts viewed the October Revolution as an illegitimate seizure of power,[56] they voluntarily cooperated with the new rulers. It is important to bear in mind that a position in a state agency was associated with an array of privileges and guarantees. Given the erosion of public order during the civil war, such privileges and guarantees became ever more important for organizing one's professional and everyday life.[57] Doiarenko, for example, owed it to his status as a 'responsible staff member in a Soviet institution' that he was able to obtain a guarantee which exempted his private flat and possessions from confiscation, along with the assurance that he could not be drafted for military service without the express consent of his employer.[58] Due to the increasing concentration of financial and administrative resources within state structures, coming to an understanding with the new rulers was also decisive for asserting interests of the larger expert community. In 1918, together with the Moscow professors V. la. Zheleznov and D. N. Prianishnikov, Doiarenko convinced the Narkomzem to grant the status of an institution of higher learning to the Higher Agricultural Courses in Saratov, which had been established as a public initiative in 1913.[59] This and many similar arrangements were rooted in a mutual pragmatic interest. Cooperation allowed the Bolshevik elite to place scientific research under state control and to utilize it for administrative purposes.[60] For the scientists it offered the possibility to institutionalize academic discourse on agriculture and agronomy and to continue their professional activities with the aid of state funding.

After the beginning of the New Economic Policy (NEP) in 1921, it seemed as if cooperation between the Bolsheviks and experts such as Doiarenko even had some programmatic grounds. The partial reintroduction of market relations, the revitalization of agricultural cooperatives and the demonstrative turn toward the peasantry were in line with the modernization agenda which the experts had been following since the turn of the century. During the time of the NEP, their professional network continued to exist within the structures of the People's Commissariat for Agriculture. The personnel makeup of Narkomzem's planning commission, *Zemplan*, was broadly similar to the pre-revolutionary conventions of experts or the agricultural advisory boards of the Provisional Government.[61] As a consequence, some pre-revolutionary traditions continued to be viable in the agrarian doctrine of the early Soviet state. Many Narkomzem publications treated the peasants as independent subjects, who in a suitable environment could become active agents of agricultural growth. This was true in particular for the first plan of the Narkomzem's activities in 1922, which was presumably written by A. V. Chaianov, and for the first general plan for the development of agriculture, which contained explicit traces of the pre-revolutionary academic debate on the viability of the peasant farm.[62] For contemporaries, it was apparently not difficult to discover a trajectory between the pre- and the post-revolutionary times. When in 1926, Doiarenko was praised as someone who had been standing twenty-five years 'face to the village', this was not only an adaptation of the dominant rhetoric of the time, when the 'face to the village' campaign was at its height. The fact that Bolshevik language helped to present Doiarenko's professional career as a harmonic whole may also be seen as an indicator that contemporaries perceived the NEP period as a continuation of the pre-revolutionary modernization programme under the aegis of the Soviet state.

However, the public reputation of the experts was not commensurate with the extensive professional competencies they enjoyed. When the socialist transformation of society was elevated to the level of official state doctrine, the pre-revolutionary image of the agronomist became disputed. In Bolshevik discourse, agronomy was not a *bona fide* profession, but rather presented a political denomination:

> The new agronomist must be a Marxist, he must be fully proficient in Marxist doctrine, and must be able to apply the method of dialectical materialism to all difficult situations in economic and in particular in agricultural life ... Economy and politics must walk inseparably hand in hand.[63]

The topos of the 'new agronomist' reflected the devaluation of the professional and moral yardsticks of the pre-revolutionary experts. These were now regarded as a self-centered ivory towered elite with a penchant for 'endless insubstantial theorizing'.[64] At the 15th party congress in December 1927, V. M. Molotov, a member of the Bolshevik leadership echelon, utilized anti-intellectual clichés to accuse the 'agrarian professors of the Narkomzem' of adhering to a line hostile to Soviet power.[65]

Stalin's rise made the discrepancy between their professional position and their lack of public recognition a real problem for agricultural experts. With the turn toward collectivization and forced industrialization, they acquired the reputation of standing in the way of the party's progressive economic policy. Because their former patrons had lost power within the party apparatus,[66] pre-revolutionary agricultural experts were increasingly exposed to harsh polemics.[67] After the reorganization of the institutional landscape of agricultural sciences and a comprehensive purge in the Narkomzem, they were finally deprived of the social spaces in which they had previously maintained a certain modicum of professional autonomy.[68] These measures aimed to prevent potential criticism of the Bolshevik leadership. When in 1929 Ia. A. Iakovlev, at that time a leading figure in agricultural policy, drafted a plan for a central academy of the agrarian sciences, he warned against the danger of making 'well-known bourgeois scholars along the lines of the Kondrat'ev-Makarov type' or the 'counter-revolutionary Professor Doiarenko' into 'Leninist academicians'.[69] Iakovlev's project suggests that, from the perspective of the Bolshevik leadership, it was self-evident that scholars such as Doiarenko had nothing in common with the Soviet project. Moreover, it demonstrates that the names of the agrarian experts had already become widely used code words to stigmatize all manner of science which was not in rigorous keeping with the party line.

The marginalization of the agrarian experts was not limited just to forcing them out of the public arena and destroying any symbolic capital that, thanks to their own institutions and high-ranking posts in the Soviet administration, they still possessed. After Stalin had laid out the direction for the collectivization of agriculture in late 1929, the experts fell victim to a change of technical and professional elites initiated by the party leadership. In 1930, leading pre-revolutionary agrarian experts were arrested under the pretext that, as the masterminds of the 'Toiling Peasants' Party' (*Trudovaia krest'ianskaia partiia*, TKP), they had been working to overthrow Soviet power.[70] This action was in keeping with Stalin's political calculations. Like the representatives of other influential groups of experts whose professional careers extended back into the pre-revolutionary period, the agrarian experts were considered a potential source of disruption in the building of socialism by dint of their visibility within the state administration and the academic sphere. Their suppression made it possible for Stalin to strengthen his own position of power and to secure the transformation of society by means of a 'new' elite which would be unconditionally loyal to the Soviet state.[71]

The Paradoxes of Soviet Clientelism

Stalinism put an end to a tradition of agricultural expertise extending over almost three decades. In the wake of collectivization, knowledge about peasant agriculture lost its relevance. Although many of the agrarian specialists arrested in 1930 were able to commute their sentence behind bars with an equivalent period of banishment, or were pardoned a few years before completion of their full sentence, they had forfeited their former status as experts. The fact that they had been accused as saboteurs of the Soviet order made it virtually impossible for them to return to their former profession. Even when they found regular employment, the doors to positions in keeping with their qualifications and many years of experience remained closed to them.[72] The social marginalization of the old elites was followed by a reign of terror. In 1937 and 1938, numerous agricultural experts fell victim to excessive state violence. Unlike the arrests of 1930, this violence was not limited to certain social milieus, but rather encompassed all strata in Soviet society. In the course of this terror, the myth of the TKP plotting against the Soviet leadership was revived. Agrarian experts and economists who had already in 1930 been accused of seeking to take over leadership positions in the TKP were once again targeted by the Soviet security authorities. The economists A. V. Chaianov, N. D. Kondrat'ev, A. A. Rybnikov, the agronomist A. V. Teitel, the cooperative activist and last agricultural minister of the Provisional

Government S. L. Maslov were sentenced to death and shot. The economist L. N. Litoshenko died at the end of the 1930s as a Gulag prisoner.[73]

Doiarenko had luck in misfortune. After his prison term ended, he was banished in 1935 to Kirov, where he became an agrotechnical consultant in the local agricultural experimental station. During his time in Kirov, he also gave lectures and classes for the staff of Machine and Tractor Stations (MTS). After such activities were prohibited to him because of his status as a banished person, he worked temporarily as a composer for a visiting puppet theater troupe in Kirov. Even after the official end of his banishment in 1939, Doiarenko remained stigmatized due to the scandal surrounding the TKP. His efforts to secure a position in various agricultural institutes and agencies dealing with agrarian policy in Moscow proved unsuccessful. After the years of terror, no one was prepared to employ a scientist who had formerly been brought to trial as a 'traitor'. Doiarenko then accepted a job as an ordinary staff member at the Grain Research Institute in Saratov, a far less prestigious institution than the Timiriazev Academy, where Doiarenko had held top-level positions until the end of the 1920s.[74]

Doiarenko's professional biography subsequent to his arrest was symptomatic of the hierarchical relationship between science and politics in the Soviet Union. A high standing within the scientific community was not a sufficient condition for employment in a scientific institute, nor did it guarantee public recognition as a scholar. Far more important was the patronage by relevant decision-makers. Doiarenko had experienced this already during the time of his banishment in Kirov. During this period, he obtained a post thanks to his personal ties with the director of the experimental agricultural station, the plant breeder N. V. Rudnitskii. Rudnitskii promised to clarify with the secret police the question of Doiarenko's employment upon the latter's arrival. Soon after that, Doiarenko started working in Rudnitskii's experimental station.[75] Similarly, Doiarenko's appointment to the Grain Research Institute in Saratov was due to assistance from the institute's director Ia. N. Itskov.[76] Even after Itskov left to take up a career in the Party Central Committee, Doiarenko enjoyed a certain degree of protection. In 1947, after he had been active for several years in the Laboratory for Agrarian Chemistry and Agricultural Soil Science, it was suggested to him that he become a candidate for the vacant chair in General Agriculture. A short time later, Doiarenko was appointed to the professorship 'as one of the most important figures in the field of agricultural sciences'.[77]

However, this reestablishment of Doiarenko's reputation as a scholar was short-lived. He fell victim to the wave of repression that engulfed the agricultural and biological sciences in the wake of the canonisation of T. D. Lysenko. Doiarenko was a critic of the soil scientist V. R. Vil'iams, who ranked among the most recognized authorities in the field of agricultural soil science at the time.[78] After the political decision on Lysenko's leading role in science, Vil'iam's crop rotation system using forage grass (*travopol'naia sistema*), which promised increased soil productivity by cultivation of multi-annual grasses, became a virtual dogma. For Doiarenko, who since the beginning of his career had championed procedures utilizing inorganic fertilizers, these developments spelled not only his demotion as a specialist for agricultural soil science, but also the end of his career. On orders from P. P. Labanov, the deputy agriculture minister of the Soviet Union, Doiarenko was dismissed from his post in 1948, along with the director of the Grain Research Institute, Ia. I. Riazanov.[79]

Just as Doiarenko's loss of reputation had come subsequent to a decision at the highest level, his rehabilitation also emerged after a signal from the political leadership. N. S. Khrushchev, General Secretary of the Communist Party since September 1953, demonstratively distanced himself from the politics of his predecessor in order to present himself as a new leader. Referring to experts, who during Stalin's reign had been stigmatized as enemies of Soviet power, Khrushchev underscored his ambitions for a change in agrarian policy. During a visit to Saratov Agricultural Institute in 1954, Khrushchev commented that Soviet agriculture would clearly be better off if specialists like N. M. Tulaikov, executed in 1938, or Doiarenko had an important say in agrarian matters.[80] In 1961, Khrushchev publicly rejected Vil'iam's dream of a blanket introduction of the grassland fallow system as untenable and counted Doiarenko among the most competent critics of Vil'iams.[81] After that, Doiarenko who had already passed away in 1958, was rehabilitated as a scholar. On the occasion of the ninetieth anniversary of the agronomist's birth in 1964, the Moscow House of Scientists organized a commemorative evening. As it had been the case when Doiarenko celebrated his professional jubilee in 1926, colleagues and former students recalled Doiarenko's merits as a scientist, pedagogue and public activist. The commemoration followed the rules of public communication in Soviet Russia. The agronomist now was again presented as a loyal supporter of the Soviet cause. One former student stressed the accordance of Doiarenko's positions with those of the Communist Party:

> The Central Committee of the CPSU and the Soviet government have rejected Vil'iams's pseudo-scientific grassland theory, finding the agronomic concepts of D. N. Prianishnikov, A. G. Doiarenko and others to be correct ... The outstanding role played by Nikita Sergeevich Khrushchev in this respect is generally well-known. For the agronomists, he is not only the First Secretary of the CPSU and the chair of the Council of Ministers, but also the most important agronomist in the USSR. And we agronomists are pleased to proceed with all speed to his assistance.[82]

Another participant of the event suggested Doiarenko as a role model for agrarian experts: 'May his shining example serve young agronomists today as a guiding star.'[83] The logic of remembrance was paradoxical: after the status of the scientist repeatedly repressed by the party had been officially restored, reference to Doiarenko became a means to assert one's loyalty to the political elite.

Modernity as a Contested Project

In this chapter I have studied the social construction of expertise in late Imperial and early Soviet Russia. Using the example of Doiarenko's professional biography I have demonstrated the mutual relations between the political, the scientific and the public spheres under different political regimes. Whether a person in early twentieth-century Russia could claim the status of an agricultural expert was decided in complex negotiations on the meaning of certain types of knowledge for solving agriculture-related problems. Moreover, in spite of a widely shared belief in the neutrality of knowledge, the outcome of these negotiations was always politically mediated. Expertise and ideology were intrinsically linked to each other: proposals to overcome the alleged backwardness of Russian agriculture contained diverse ideas about what the future village should look like. These ideas, on their part, condensed certain visions of the social and political order on a more general scale. Hence, the authority and prestige of agricultural experts at different times is also a lens through which we can explore how contemporary actors negotiated Russia's future.

Examination of the career tracks of experts such as Doiarenko allows reflection on the role of early twentieth-century Russia in modern history from a more general perspective. It has become common to integrate Russia into a shared history of modernity, regardless of the political system and ideology which seemed to distinguish Imperial Russia and the Soviet Union from the west. A powerful argument in favour of that idea is the fact that some basic processes of modernity such as scientification, bureaucratization, nationalization and industrialization in Russia as in other parts of the world coincided with projects to transform the world in accordance with science and reason.[84] Following this tradition, the Stalinist state and Soviet collectivized agriculture have been regarded as the embodiment of an authoritarian high-modernity.[85]

The case of Doiarenko clearly confirms Russia's participation in the project of modernity, helping at the same time to interpret it in a more differentiated way. Focusing on peasant households, cooperatives and self-government, experts such as Doiarenko developed visions and methods which differed from the radical ideas and authoritarian practices which are often associated with Russian and Soviet modernity. The fact that in their work many agricultural specialists focused on the peasant family as the main social and economic unit in the countryside has been interpreted as an indicator of their allegedly anti-modernist spirit. Contemporaries, as well as historians who adhered to modernization theory, were convinced that members of the expert elite idealized a pre-modern, stationary economy and thought of the village as the last bastion of a past golden age.[86] However, the vision of gradual rural development, which was the cornerstone of Doiarenko's perception of himself as an expert was not rooted in a discomfort with change and development. Instead, it was based on the belief in an alternative future. The case of Doiarenko's professional biography demonstrates that in Russia as elsewhere different projects of modernity competed with each other. Whether a person was effective in claiming to be an expert depended on the dominant vision of modernity at any given time. Expertise was as contested as was modernity itself.

Translated from the German by Bill Templer.

7 THE RISE OF THE SCIENTIST-DIPLOMAT WITHIN BRITISH ATOMIC ENERGY, 1945–55

Martin Theaker

The growing influence of science in national policymaking structures during the twentieth century has been a well-documented phenomenon. Throughout the period, scientists and other forms of experts gained increasing significance within government procedures, in turn stimulating debate regarding their role in democratic systems. However, this trend did not occur universally or even uniformly, making it necessary to particularize the figure of the expert by identifying traits specific to certain specialist groups in their appropriate geographic and temporal contexts.

In the last century, few scientific discoveries proved as important as that of atomic fission, and Britain's role in establishing this domain, alongside the subsequent assumption of the pioneer's mantle by the United States, is widely understood. British scientists played a crucial role in the Allied atomic bomb project, gaining considerable experience from their participation in large-scale projects at Montreal and Los Alamos. This small cadre of elite scientists, many of whom were world leaders, thus represented an incredibly valuable resource, and their importance was magnified both by their scarcity and the breakdown of Anglo-American information-sharing agreements in 1946. Consequently, on returning home, such men soon became indispensable to post-war governments hoping to exploit atomic energy as a solution to Britain's serious economic and geopolitical problems. In this way, obtaining an independent nuclear weapon and developing civil atomic technology to meet projected imperatives in the energy economy became both a political priority and an area of rich scientific potential in the United Kingdom.

However, an aspect of these events that has hitherto lacked appreciation is the species of expert that evolved within this unique context. Developing atomic energy for civil purposes in post-war Britain quickly proved problematic, as the nascent enterprise lacked manpower, resources and finance. It also presented governments with the paradox of arranging national scientific talents into structures which would preserve the freedom to research whilst maintaining firm political control over their objectives. The main purpose of this chapter is therefore to analyse the role of the individuals who assumed such great influence and to evaluate their effectiveness within British political and scientific processes. The 'scientist-diplomat', as he has recently been represented by Andrew Brown, is identified as an agent of interaction between Britain and foreign states.[1] This type of expert, preponderantly an atomic physicist, has been widely discussed by scholars for his role in navigating governmental machinery to conduct international scientific relations, particularly in relation to European organizations like CERN.[2] This analysis, however, will expand this definition by contending that the phenomenon had deeper roots within *domestic* organizational changes. In this sense 'diplomacy', defined traditionally as interaction between foreign powers, is worthy of examination as an exchange between the two separate *worlds* of the scientific and the political. This is a rich scholarly field, as the vein of biographies currently being tapped clearly demonstrates; in addition to Brown's study of James Chadwick, Sabine Lee has examined the correspondence of Rudolf Peierls, whilst Peter Hore has investigated the life of Patrick Blackett.[3] Furthermore, characters such as Hans Bethe, Edward Teller and Robert Oppenheimer continue to fascinate American scholars, building a broad foundation of personal stories surrounding atomic physicists.

The role of technocracy in the modern state has also produced fruitful debate. In her work, Sheila Jasanoff highlighted a conflict between the 'democratic' and 'technocratic' interpretations of science policy, with proponents of the latter demanding the integration of 'more and better science into decisions' to guarantee successful outcomes.[4] In order to test this notion of beneficially increasing the quality and quantity of scientific input, this chapter will investigate the role of the expert in a scientific field which commanded considerable priority in post-war British policy. During the late 1940s, civil atomic energy emerged from the shadow of military applications to attain substantial significance in its own right, forcing the construction of novel frameworks to accommodate its demands. Within this state of flux, formative issues raised themselves, including the organization of scientists working on nationally-important projects and the degree of government control to which they should be constrained. Therefore, this chapter will discuss how the scientist-diplomat evolved within the political and scientific circumstances specific to post-war Britain, and particularize the figure of the expert in this context.

Beginning with the background to physics research in the interwar period, this chapter will therefore locate the beginnings of atomic energy in the UK in the Allied weapons programme of the Second World War. The economic and military reasons for atomic power's overarching importance will then be established, highlighting the rapidly-accelerated influence of atomic energy experts in Britain. The main focus of the analysis will consider the construction of organizations designed to direct atomic energy research in Britain, and the role of scientists and politicians within these frameworks, before assessing how expert influence produced a change in organizational structures to accommodate the changing relationship between science and state. At this stage, the motivations of the scientists themselves will be evaluated and their position as elites within a new industry based around sensitive technology considered. Finally, this article will conclude by discussing the issues raised by Britain's atomic restructuring at the interface of science, industry and government, including the contradictions inherent in its increased international role. In this way, a sharper definition of technocracy in this distinct environment will be produced, and a greater understanding of the evolving role and effectiveness of the expert in the political and economic milieu specific to Britain in the 1950s ascertained.

The Second World War and the Atomic Bomb

The early twentieth century was a period of intense scientific and engineering progress. Stimulated in part by the Great War, new technologies emerged, notably in the automotive, chemical and aeronautical sectors. In Britain, governments had already made notable efforts to encourage the growth of modern science, and to control it via state machinery. In 1902, the National Physical Laboratory opened in Teddington amid a spate of university-building which saw five civic 'redbrick' institutions founded for the purposes of promoting engineering and other practical disciplines. Concurrently, the Department of Scientific and Industrial Research was founded in 1915 to coordinate government funding to university and private industrial research. Academic institutions provided the springboard for what the eminent physicist John Cockcroft would later describe as the 'Renaissance of Physics', which occurred in the first decades of the twentieth century.[5] During this period, British-based scientists built a substantial grounding in atomic physics at university level, particularly in Cambridge, where Chadwick's discovery of the neutron, and Cockcroft and Ernest Walton's 'splitting' of the atom (both in 1932) both occurred. The Cavendish Laboratory acted as a nursery for several of the greatest talents in the history of British physics, representing a common alma mater for numerous physicists involved in the development of atomic energy in the United Kingdom. As the 1930s drew on, these domestic talents were complemented by numerous European *émigré* scientists fleeing the rise of National Socialism. Among them were Otto Frisch and Peierls, two theoretical physicists from Vienna and Berlin respectively, who settled at the University of Birmingham. Significantly, therefore, Britain's physics cohort consisted of foreign experts complementing domestic talents, who often enjoyed shared educational backgrounds, with both groups organized into local centres via research posts at prominent universities.

The outbreak of war focused atomic research on the potential military applications of nuclear fission, and in March 1940 Peierls and Frisch devised their famous memorandum which estimated that the critical mass of Uranium-235 required to produce an atomic bomb was as little as one pound (approximately 450 g). The work was forwarded to Henry Tizard, chairman of the Aeronautical Research Committee, who immediately formed the 'MAUD Committee' to discuss the practicality of a British atomic weapon. The group comprised chairman George Paget Thomson and his Cambridge colleague Cockcroft, along with Chadwick from Liverpool University, the Birmingham scientists Marcus Oliphant and Philip Moon, as well as Manchester physicist Blackett. Ironically, Peierls and Frisch, as 'enemy aliens', were (temporarily) excluded from advancing their own work via the committee.[6] Nonetheless, although British experts had acquired the *theoretical* knowledge to produce an atomic weapon, limited resources and engineering capacity, coupled with the obvious urgency of obtaining such a bomb before the Axis powers, encouraged Whitehall to negotiate with its American allies about combining atomic programmes. Accordingly, the Quebec Agreement of 1943 declared that in the 'wise division of war effort' Britain would merge its atomic research, codenamed 'Tube Alloys', into the Manhattan Project in exchange for all information derived from the venture. However, the Agreement sowed the seeds for future strife in its fourth clause, which stipulated that 'the Prime Minister expressly disclaims any interest in industrial and commercial aspects beyond what may be considered by the President of the United States to be fair and just and in harmony with the economic welfare of the world'.[7]

Subsequently, eminent British scientists were arranged into important projects on either side of the Atlantic. Cockcroft developed radar technology in Malvern before assuming the directorship of the Canadian Atomic Energy Project at Chalk River in 1944. Blackett moved into operational research, devising bombing strategies for Allied aircraft, while Chadwick, Peierls and Frisch (the latter pair now trusted) departed for the Manhattan Project. An important exception to this trend was the Oxford physicist Frederick Lindemann, known more commonly as Lord Cherwell, who became Churchill's personal scientific adviser during the war years. As a noted mathematician, Cherwell involved himself with economic strategy from his annexe at 10 Downing Street, and in so doing represented a unique facet in the way expert opinion was relayed during the pre-institutional era, with his personal advice to Churchill culminating in an appointment as paymaster-general to the wartime Cabinet in 1942.[8] However, by entering the political stream, Cherwell bound his success and influence to the fortunes of the Conservative Party and in particular, Churchill.

Tube Alloys and the Manhattan Project were therefore instrumental in the development of state-science relations, because the extreme circumstances of war forced government to harvest domestic physics talent, propelling scientists into 'national' work with political goals. Naturally, scientists had co-operated with the government prior to 1940, but Britain's undeniable defence needs during the period compelled many scientists into fields such as atomic energy, radar and bombing strategy with far greater urgency. Tube Alloys can thus best be viewed as a project displaying an *abnormal* national concentration of physics talent, establishing the principle of state coordination of atomic energy specialists on a justifiable ethical basis, thereby laying the foundations for future large-scale physics projects.

The Ethics of the Bomb

Yet such assemblages were also significant in their implications for the effectiveness of scientists to communicate the moral qualms inherent in their work. Indeed, the University of London physicist (and Cockcroft's erstwhile Cavendish colleague) Harrie Massey identified the roots of a collective scientific conscience regarding atomic energy in the 'Big Science' scale of the undertaking.[9] Generally speaking, scientific experts accepted the case for a British atomic bomb, motivated, particularly during the early years, by the fear that German scientists would get there first. However, it would be misleading to conclude that all British (or indeed Allied) scientists were therefore permanently favourable to building or even researching atomic weapons. Perhaps the most famous case of dissension from 'national' projects was the resignation of Joseph Rotblat, a Polish-born British physicist, from the Manhattan Project in late 1944. A firm believer in scientific ethics, Rotblat felt disgust at potentially enabling a future American atomic strike on the USSR, when Soviet soldiers were dying in droves pursuing the defeat of Nazi Germany that the bomb project was supposedly hastening. Importantly, however, he identified three major motivations among those scientists who chose to continue even after it became clear that Germany could not obtain nuclear capabilities. Firstly, intellectual curiosity propelled many to see if the bomb *could* be built, while others feared for their career prospects if they wavered. Additionally, Rotblat contended, some scientists believed that atomic warfare would be used legitimately to compel Japan's surrender, although he designated this last group a minority, with most, he opined, not being 'bothered by moral scruples'.[10]

Albeit an isolated case, Rotblat's departure coincided with a groundswell of concern which manifested itself in the formation of scientific groups distinguishable into two main types. The first route was that taken by the founders of the Association of Scientific Workers (AScW), who aimed to create a trade union for scientists and to comment on political affairs. The group had supported the implementation of scientists in Britain's defence, citing the threat of fascism as an inherently anti-scientific and anti-democratic evil.[11] However, their aims were fundamentally pacifistic, being concerned with increasing the representation of scientific ethics in national and international policymaking. Blackett became president of the association from 1943–6, although his socialist views did not necessarily guarantee his support of the Labour Party as some peers assumed.[12] The second direction was that chosen by the Atomic Scientists'

Association, spurred by Rotblat and Peierls, which aimed to educate the public and politicians whilst advocating international control of atomic power.[13] These associations mirrored international trends which culminated in the famous 1955 manifesto issued by Bertrand Russell and Albert Einstein beseeching global governments to forsake atomic weapons in conflict resolution.[14] In turn, such calls led to gatherings like the annual Pugwash Conferences, which began in 1957 with the aim of reducing international tension via scientific discussion.

Despite its comparatively small scale, the opting-out of scientists from the atomic weapons programme marked the first dissension of experts from government will. In this extreme example, the limits of democratic government to coerce scientists, even in times of war, were clearly demonstrated. For their part, the feelings of scientists regarding their technology were informed by numerous factors, including political stance, national identity, personal background and in some cases, strategic concerns. Therefore, in their role as experts many scientists felt obliged to exercise *indirect* control via the moral arguments they imposed, laying the foundations for the role of the scientist-diplomat as a reformer as well as an adviser.

The Need for British Atomic Power

Ultimately, the Anglo-American atomic effort came to fruition in August 1945. At a stroke, the bombings of Hiroshima and Nagasaki changed the international political scene and drew into question the ethics of a technological field in which scientists had played a central role. Nevertheless, although the British scientific mission undoubtedly contained highly-skilled scientists, these experts had merely participated in a large project with a defined objective, and at that, one based abroad. Establishing a permanent *technocracy* in the UK, at least with regard to civil atomic energy, would prove more challenging. The general election of July 1945 returned a Labour government promising to keep a 'firm public hand on industry' to secure full employment.[15] This thirst for social reform stretched across diverse areas of national life: railways were nationalized, a centralized health service established and the welfare state significantly expanded to improve pensions and working conditions. Labour's socialist ideology promoted centralized government and, as a national security concern, atomic energy in the UK was therefore doubly consigned to begin life under state control.

There were also important diplomatic issues to be resolved. By merging their atomic effort into the American programme, the British had guaranteed the swifter implementation of the product, but had also exposed themselves to protracted political negotiations regarding the future of the technology amid a fluctuating international scene. In November 1945, British Prime Minister Clement Attlee, US President Harry Truman and Canadian Prime Minister William Mackenzie King issued their famous 'Washington Declaration' which proclaimed the need for an international control agency for atomic weapons under the United Nations. After lengthy debate, the three atomic partners also underlined their willingness to assist the development of civil atomic energy, but stipulated that this could occur only once atomic weapons were adequately safeguarded.[16] Simultaneously, however, Attlee quietly dispatched Sir John Anderson to negotiate Britain's release from clause four of the Quebec Agreement, which met with success after Anderson ceded Britain's rights to large uranium stockpiles.[17] Such hopes of continued wartime collaboration were short-lived, however, and officials in Whitehall soon found themselves rebuffed by influential senators who feared nuclear proliferation outside the United States. By August 1946, Truman had been compelled into signing the McMahon Act, effectively ending all Atlantic co-operation on atomic energy. This rupture was crucial, because the Act essentially expelled foreign scientists from the United States, creating a cadre of British physicists returning from wartime projects without large-scale domestic institutions in which to continue their research. Simultaneously, the government acknowledged that atomic experts were vital in addressing two national issues: weapon production and civil energy generation.

The first of these imperatives was utterly non-negotiable: the atomic bombings of Hiroshima and Nagasaki had caused massive casualties with merely two bombs and the potential for similar havoc in the densely-populated UK was feared. Possessing nuclear weapons was therefore considered key for continued international recognition in the Cold War, as demonstrated succinctly in autumn 1946 at a meeting of Attlee's 'GEN-75' group dedicated to investigating the possibility of a British nuclear deterrent. Sensing the committee opposed the idea, Ernest Bevin, the secretary of state for foreign affairs, exclaimed 'we've got to have this thing over here whatever it costs! We've got to have the bloody Union Jack on top of it!'[18] Many influential scientists concurred, with Cherwell arguing that without atomic weapons, Britain would be 'in the position of savages armed with boomerangs and bows and arrows confronting armies using machine guns', while Chadwick believed the bomb would render large-scale conflict unthinkable.[19]

The second important application of atomic energy lay in civil electricity generation. After an initial post-war fillip, British coal production plateaued in the early 1950s, whilst the proportion of output used in domestic power plants rose consistently. As a result, exports crumbled, decreasing British trading influence and setting a collision course for an energy gap.[20] Brutal winter weather in 1946–7 forced many coal-fired plants to close due to lack of supply, further highlighting the weakness of Britain's fuel infrastructure. Simultaneously, the carbon alternative, oil, was becoming unattractive due to import cost and its propensity to make Britain dependent on the Middle East, a region demonstrating increasingly anti-imperialist tendencies. Oil also presented a geopolitical problem: as David Painter has highlighted, oil was incorporated into a policy enacted by the

United States via the Marshall Plan to make western Europe dependent on American fuel companies.[21] Although in its absolute infancy, atomic energy presented a novel solution to both of these problems, and as early as 1950 British engineers had presented semi-detailed calculations observing that the technology could become competitive with coal-fired generating plant in the near future.[22]

In terms of British technocracy, therefore, the most important aspect of these imperatives was that atomic engineering elevated energy solutions above the 'lay' stratum. The management of fossil fuels could be comprehended by political classes without specialist education, particularly within established fields of government activity such as labour relations or import policies, but atomic energy, by contrast, necessitated agency from highly-educated elites who were suddenly propelled into a position of serious national importance. At a stroke, the politician was reduced to an administrator, overseeing and directing (where possible) the activity of physicists, chemists and engineers whose work he could hardly be expected to understand. Thus, the extent to which these imperatives should allow governments, in league with experts, to obviate democratic processes was inherently bound to the development of atomic energy in the UK.

The Committee System

Put simply, the Labour government faced the problem of cultivating atomic technology for dual purposes with minimal resources outside the intellectual field. In the interests of continuity, the wartime system of committee organization was prolonged, with an 'Advisory Committee on Atomic Energy' (ACAE) being established in August 1945 under Anderson, the Conservative Cabinet member who had been responsible for British atomic energy during the war.[23] It included among its members a balanced representation of the most suitable individuals and institutions available in British science and government, including scientists, Foreign Office officials and members of the Royal Society. Appended to the ACAE was a 'nuclear physics sub-committee' (NPC) comprised purely of physicists, notably Chadwick, Peierls, Cockcroft and Blackett, which was tasked with 'making recommendations regarding the programme of nuclear physics to be pursued in the country as a whole'.[24] Clearly visible in this committee structure was an attempt to blend expert opinion with military and civil service interests alongside traditional institutions, like the Royal Society, which claimed to represent authentic scientific opinion. Nevertheless, this approach was unpopular with certain groups, notably thirty-one scientists in the Montreal Group who questioned why Anderson should be allowed to continue when the Cabinet he had served had just been 'rejected by the whole nation'. Instead, the group requested that an active Cabinet member chair the committee and that the director of any future national atomic energy establishment 'should be a man of outstanding research experience in nuclear physics, not an administrator steeped in Civil Service tradition' who would in any case answer only to a minister.[25] Furthermore, the system of personal invitations was treated with suspicion in some elite quarters, and Oliphant wrote to Blackett early in 1946 to voice his concern that science was becoming too intimate with the politics of the Labour government. He relayed a comment from Herbert Skinner (later head of general physics at Harwell) insinuating that unsatisfactory research programmes were the result of 'yes-men' rather than 'honest-to-goodness scientific men' being chosen to advise government.[26]

Although susceptible to suspicion regarding the validity of individual experts to represent a spectrum of scientific opinion, the committee system was employed widely by government throughout the atomic energy field, notably at the interfaces where the technology coalesced with defence interests. Such influence could also be wielded on a smaller scale by one expert; beginning with the Ministry of Defence in 1948, government departments increasingly procured in-house expert opinion by appointing chief scientific advisers to assist decision-making. In a specifically atomic context, however, this constant interaction with politicians and administrative systems stimulated the evolution of a new, politically-savvy breed of expert, namely the 'scientist-diplomat'. This term, used by Andrew Brown in his biography of Chadwick, depicts a figure who achieved prominence at the point where traditional scientific channels required representation to government in order to negotiate greater autonomy or influence.[27] As a priority field, atomic energy therefore allowed leading physicists, as scientist-diplomats, to somewhat increase their leverage during this period of fluid organization when concrete policies remained undecided. Perhaps obviously, however, this new effectiveness was not unconstrained, and although physicists were bound by common backgrounds or overarching ethics, the committee system demonstrated that they were themselves still 'atomised' in the face of government. Instead, experts were arranged within the orbits of prominent individuals in the wider 'science network', including politicians, civil servants and military brass. This was partly a result of the ad hoc organization forced by war; Cherwell's advisory role to Churchill was largely unofficial and after 1945 the ACAE continued the wartime committee system via appointments made on Attlee's personal authority as prime minister. During and immediately after the war, the strongest links between many of Britain's atomic energy experts and government were therefore personal relationships conducted unofficially outside any atomic establishment. Soon though, advancing 'Big Science' effectively would require larger and better-defined institutions.

The Atomic Energy Research Establishment

To pioneer atomic energy research, a government-controlled establishment would need to recruit and organize experts currently based at universities or abroad. As stated previously, disharmony with Washington, along with security considerations had rendered the British research effort utterly dependent on domestic (and Commonwealth) manpower, greatly magnifying the leverage of these scarce individuals. Furthermore, British scientists returning from North America, among them Cockcroft, Peierls and Chadwick, enjoyed a sizeable advantage in having already witnessed the results of Allied research and development, allowing them to avoid duplicating existing engineering studies. In any case, although no decision had yet been taken to produce an atomic bomb or to enact a civil nuclear power programme, both applications were appreciated in Whitehall, even if officials did not immediately foresee when they were to be implemented. Therefore, the prime minister proposed a domestic programme to research atomic technology and produce fissile material 'as circumstances might require', consequently marking the tentative but significant theoretical separation of peaceful atomic energy from weapons development.[28]

Security considerations and a centralising ideology led the Labour government to form the Atomic Energy Research Establishment (AERE) at Harwell in 1946 under the direct control of the Ministry of Supply, with Cockcroft as director. Initially, the AERE conducted basic research, provided information to the Industrial Group and produced radioactive and stable isotopes for industrial and medical uses.[29] The Industrial Group, charged with producing plutonium, was established under Christopher Hinton at Risley, while William Penney returned from America to head the 'High Explosives Research' project, both also under the Ministry of Supply.[30] From the start, Cockcroft's new organization faced three core challenges. Firstly, there was the issue of establishing an autonomous atomic programme against the grain of scientific internationalism in a field which had in its short lifetime already generated a web of national claims and interests. British wartime atomic research had been assisted by escapee French physicists, while Belgium and Portugal had both signed uranium contracts.[31] Most important, however, was the need to extract rights from the Allied project when co-operation with Washington was frozen and new partnerships were proscribed by declassification agreements with the USA and Canada.[32] Some gains in this regard were made through the 1947 'Modus Vivendi', an unofficial agreement under which Britain released hoarded uranium to the United States in exchange for additional atomic knowledge. Indeed, these negotiations provided a notable demonstration of the importance of experts, as eminent scientist-diplomats including Cockcroft managed to temporarily revive the Anglo-American atomic relationship where politicians had failed.[33] Nonetheless it was a minor fillip, and such tangled commitments ensured that Britain's militarily and commercially sensitive project would simultaneously be expensively independent and dogged by frequent international renegotiation.

A second dilemma was that of establishing an atomic programme with limited resources during a period of severe austerity. The situation was eased somewhat by designating atomic energy a priority field, and both the Labour and Conservative governments increased atomic energy expenditure substantially year-on-year during this period.[34] Nonetheless, a huge effort deploying the domestic intellectual capital of the scientist-diplomats was required, and so the final concern for government lay in organizing such a large project, representing as it did Britain's first foray into state-funded 'Big Science'. The challenge was to construct an institution under national control which could concentrate scientific expertise and achieve research goals whilst advising government reliably. This would demand a novel approach distinct from any previous political attempt to incorporate science into policymaking, because atomic energy crossed a significant threshold in requiring national support for a scientific endeavour, making it unclear whether traditional relations between science and state could survive the intense pressure caused by its demands. Accordingly, the conditions of service offered to the scientist-diplomats are noteworthy, with Margaret Gowing observing that Cockcroft was given relatively extensive powers to determine scientific activities at Harwell, a view supported by the terms of his appointment, which promised him conditions similar to those in a university research laboratory.[35] Additionally, Cockcroft, Penney and Hinton, along with ministerial administrators, formed an Atomic Energy Council which would decide policy, under the chairmanship of Lord Portal, the new Controller of Production (Atomic Energy) at the Ministry of Supply.[36] The three pre-eminent 'barons' of Britain's atomic energy effort were therefore granted considerable influence at the policymaking and research level, albeit still within direct government confines.

Subsequently, the responsible legislation, the 1946 Atomic Energy Act (distinct from its American namesake, the 'McMahon Act') invested in the minister of supply the power to 'promote and control the development of atomic energy' with funding directly from Parliament.[37] The Act itself was adopted in a passionless debate amid cross-party consensus, punctuated only by calls from the ASA to loosen security restrictions.[38] Such concerns were supported by letters in the national press, and in November 1946, Peierls and his Birmingham colleague Philip Moon wrote to the *Times*, expressing their disappointment that 'government could not agree to maintain scientific freedom as a legal right rather than by ministerial order'.[39] Indeed, the interface between this keen body of scientific conscience and 'institutional science' was initially potent with conflict, as atomic scientists, in their role as civil servants, felt discouraged from participating in scientific associations. Cockcroft adopted the cause, and after difficult negotiations, successfully persuaded the Ministry of Supply to allow AERE staff to join the ASA in order to moderate the association and prevent its domination by left-wing scientists.[40] Thus, the scientist-diplomats negotiated a compromise between the two groups, highlighting both their growing skill in negotiating with administrators and rising profile within departmental processes.

Nonetheless, forming atomic institutions did not signify a break with existing governmental methods. Having fulfilled its original purpose, the ACAE was disbanded in 1948, but far from being superseded, the committee system itself was allowed to multiply with four new committees assuming the ACAE's duties. Of these, the Ministerial Committee and the Official Committee on Atomic Energy were each populated exclusively by ministers and civil servants, but the new Production Committee and Defence Research Policy Committee both recruited Chadwick, now back in Liverpool, to represent scientific interests.[41] Broader atomic energy questions could also be referred to the Advisory Council on Scientific Policy.[42] Importantly, therefore, the 1946 Act represented an attempt by the political elite to accommodate atomic energy within the traditional organizational structures it understood, forcing national science to perform within a departmental straitjacket. Within this framework, atomic scientists were essentially treated as civil servants staffing the department of atomic energy, causing the glamorous new technology to suffer almost immediately from a shortage of manpower once civil service salaries failed to match what private industry could offer promising physics and engineering graduates.

Scientific Motivations

As a 'Jekyll and Hyde' technology, atomic energy exposed varying motivational patterns among scientists, with many preferring to remain in their university posts rather than join the centralized establishments. Certain individuals therefore merit brief biographical analysis to distinguish particular subdivisions within the scientist-diplomat strain. To an extent, the Second World War had submerged the ethical debate over atomic weapons beneath the Manichean dialectic of 'good versus evil', because scientists could justify their work as critical in stopping fascism. For many, advancing the peaceful aspects of atomic energy presented a considerable opportunity to redress the emphasis within a previously military technology. Cockcroft in particular accepted the duality of his field but targeted a better future by directing a pioneering civil power programme and enthusiastically promoting peaceful atomic energy.[43] Uranium fission could theoretically liberate energy sufficient to enable previously unimaginable engineering schemes, and by the mid-1950s, Cockcroft was encountering suggestions that atomic energy could legitimate megaprojects such as the irrigation of semi-arid areas of South Africa by draining swamps in Bechuanaland.[44] Although fanciful, such suggestions highlighted starkly a sense of excitement among scientists and a major motivation propelling experts to develop atomic technology.

In a similar vein, Blackett saw in atomic energy a method for lifting millions in newly-independent India out of poverty whilst simultaneously discouraging weapon proliferation on the subcontinent. In this sense, he pursued the role of the scientist as an agent in political affairs, continuing the trend he had followed with the AScW.[45] Notably, Blackett himself was not directly connected with civil or military atomic energy, believing the latter unnecessary and dangerous. Instead, he used his voice within the NPC to coerce his fellow scientist-diplomats, and when this failed, he published his thoughts on atomic warfare widely.[46] In his 1948 work *Military and Political Consequences of Atomic Energy,* Blackett contended that an arms race was politically futile and highlighted the impossible position of the atomic scientist who watched forlornly as his research was used by government 'not so much to end a second world war as to inaugurate a third cold one'.[47] Men like Peierls and Frisch, who had toiled on atomic weapons within the context of a 'race' against fascism were similarly moved to develop peaceful atomic energy, with Peierls in particular keenly representing a scientific conscience to the public. Accordingly, despite returning to Birmingham, Peierls remained attached to the AERE for consultancy work after the war.

Such attitudes, therefore, are significant in demonstrating how scientist-diplomats were above the 'conventional politics', which lost relevance when developing a technology which itself raised issues on a higher humanitarian plane. Of these scientists' patriotism there can be no doubt: Cockcroft and Blackett, although pacifists at heart, served, in varying capacities, in both world wars. Indeed, Cockcroft remained on the atomic energy project despite serious offers to transfer elsewhere, believing it the best method of representing science and affecting change from the inside.[48] Consequently, through the experiences of these men a deep conflict in the role of the scientist-diplomat can be identified: how could they marry the idea of developing science for the good of mankind with the national frameworks provided in Britain? In response, some experts maintained their distance from the 'Big Science' establishments, preferring instead to assume power through traditional channels. Cherwell, now returned to Oxford, was convinced of Britain's need for atomic weapons and maintained a consultancy position at the Ministry of Supply to exert influence wherever possible. He also remained critical of overt state control and used personal contacts to press his case even during his time out of direct power, notably imploring Blackett in May 1946 to persuade the ACAE to streamline the bureaucracy that he considered 'deleterious to progress'.[49] On resuming his Cabinet position in 1951, Cherwell continued to be motivated by duty and personal loyalty to Churchill, even though the resultant overwork often damaged his health. His belief in the need for an independent board for atomic energy consumed much of his time, but crucially gave those scientist-diplomats in favour of greater autonomy a 'man on the inside' at Whitehall. Thus, the conditions were primed for the construction of an effective, influential technocracy magnified by the need for atomic technology and the scarcity of the experts themselves.

The Role of Science in the Modern State

This rising scientific influence in research issues engendered dissatisfaction in some elite physicists with their institutional frameworks. Progress slowed as every decision had to be ratified in the time-honoured civil service manner before ultimately requiring official Treasury sanction. Many scientists became disillusioned, notably Peierls, who dispatched a lengthy letter to Chadwick in May 1946 detailing the problems arising in the first few months at Harwell. Peierls implored Chadwick and Cockcroft, as the 'people with the best knowledge of policy and contacts at a higher level', to demand more efficient scientific organization, and threatened to strengthen the scientists' hand in relations with government by organizing mass resignations if the Ministry of Supply did not reduce inertia.[50] Importantly, therefore, Peierls identified at an early stage two of the most prominent scientist-diplomats and the mechanisms they might employ to catalyse changes benefitting scientists in the new atomic sector.

Ultimately, change came in 1951, when Churchill returned as prime minister, bringing with him Cherwell, who resumed his advisory role as paymaster-general. Barely three weeks after winning the election, Churchill wrote to Duncan Sandys, his minister for supply, to inform him that atomic energy would soon be removed from his portfolio and to report to Cherwell, whom he was establishing as an unofficial consultant with access to all atomic personnel, information and establishments.[51] Whilst this transitional arrangement was in place, the Conservatives continued Labour's centralized control, believing that Britain's atomic goals were too crucial to justify organizational tampering. In due course, these targets were achieved: Britain detonated its first atomic bomb in 1952 and began a programme for electricity production the following year. Nonetheless, the fact that the USSR had beaten Britain in developing atomic weapons caused alarm, bolstering Cherwell's vehement criticisms that strict departmental organization was simply inappropriate for the job. Change, he argued, was imperative to 'regain the place in nuclear development to which the outstanding achievements of our scientists entitle us'.[52] Cockcroft too adopted the cause, personally convincing Lord Citrine, the chairman of the newly-nationalized British Electricity Authority, that atomic energy would be vital to national development.[53] In September 1952, therefore, Cherwell circulated his proposals for an independent atomic corporation to the Cabinet and defended them at subsequent meetings. He stressed the fact that Britain had finally obtained an atomic weapon, negating the retarding effects of organizational upheaval, and noted that the growing civil atomic energy programme would soon require a form of leadership more akin to an industrial concern.[54]

Unsurprisingly, stiff opposition was raised by Sandys, who contended that there was no evidence that atomic research had 'been held up by the dead hand of the Civil Service' and that a new corporation would not gain real independence if it remained dependent on public finance.[55] Nonetheless, other Cabinet members were supportive, with Oliver Lyttelton, the secretary of state for the colonies (and former industrialist), highlighting that departmental organization did not encourage scientists to pursue research with complete freedom.[56] He also noted the impossibility of raising salaries in the atomic sector to attract the best industrial talents without simultaneously unbalancing the uniformity of pay across the civil service. If atomic energy were kept under the Ministry of Supply while conditions were improved, he argued, 'disorder would follow and civil disturbances would soon be detected in the Athenaeum Club'.[57] Buoyed by this support, Cherwell restated his claim in a Cabinet discussion in early November and received further backing from several ministers, notably the foreign secretary, with only the Treasury hesitant.[58] The political battle was won; the majority of politicians, with growing cross-party support, now appreciated the benefits of civil atomic power and supported the scientists' request for autonomy.[59]

Finally, the issue was put before an independent committee under Anderson (the 'Waverley Committee') in 1953, which identified that the three atomic establishments communicated ineffectively, with their respective heads meeting only to confer as individual representatives rather than assuming collective responsibility. As a solution, the panel recommended a unified board to combine the three 'legs' of the atomic energy effort. The three directors would retain the autonomy over their research, but would now assume greater administrative duties by manning a board serving a chairman who would report to Cabinet. This compromise was symptomatic of the rising influence of science; the civil service had been unable to manage scientific activities effectively and so this ideological concept was abandoned in favour of removing administrative interest to a higher level.[60] No longer would management involve day-to-day communication between scientists and civil servants at the Department of Atomic Energy.

The United Kingdom Atomic Energy Authority

The recommendations of the Waverley Report were broadly accepted, and the United Kingdom Atomic Energy Authority (UKAEA) was founded in July 1954. A board system was implemented, with the three leaders of the AERE, Risley and AWRE respectively being promoted to full membership capacity under the chairmanship of Edwin Plowden, a noted administrator with Treasury experience. In addition, three part-time members were appointed, comprising a trade unionist, an industrialist and Cherwell. Ultimate responsibility for the direction of the Authority, including the appointment of the board, was transferred to Lord Salisbury, the lord president of the council. However, the Act stated that 'no such direction shall be given *except after consultation with the Authority*, and the Lord President *shall not regard it as his duty to intervene in detail* in the conduct by the Authority of their affairs' (author's emphasis).[61] Additionally, financial provisions were altered so that funds voted by Parliament were assigned as a grant-in-aid to be allocated as necessary by the lord president, rather than administered single-handedly by the minister of supply. Thus, the new system would reduce overt government control, with finance and official policy filtered indirectly through a Cabinet member who was required to negotiate with the board at all times. In essence, the Authority constructed a bottom layer of scientific experts, whose performance was channelled through the chairman's industrialist-civil service stratum, itself acting as an intermediate to the upper layer of Cabinet and Treasury consent.

The actual scientific and engineering activities of the Authority would be based around those men whom official historian Gowing auded the 'outstanding leaders' of Britain's atomic energy effort.[62] Their talents were certainly highly prized by government, and the trio were invited to negotiate the terms of their employment, salary and pensions before being offered memberships.[63] By incorporating a steering board, a concession was therefore made to greater expert influence in policy and the Authority marked a watershed in defining the role of science within the state. However, the combination of the research, industrial and weapons groups generated conflict between scientists and industrialists regarding traditional scientific universalism. In a memorandum to Plowden, Hinton wrote that 'Cockcroft, with the majority of members of the Board, takes the view that information should be as widely disseminated as possible, and that the more information is spread, the faster will the bounds of knowledge be advanced'. Defending his position, he contended 'that knowledge is of value only to the extent that it can be used and that it is not worthwhile to pay for any knowledge which cannot be put into use. This is the industrial point of view'.[64] Consequently, Hinton argued that the Industrial Group had actually preferred to renegotiate control mechanisms *inside* the Ministry of Supply and had loyally submitted to change only in the national interest. This was the crux of the problem: Britain's scientists had achieved greater autonomy from government by restructuring as an autonomous corporation, but acting in a manner associated with an industry often contradicted co-operation and freedom of communication. For example, new industrial achievements raised the potential for exports and international negotiation, which would ultimately have to be processed at Foreign Office-level. Additionally, the administrative burden of running an autonomous organization soon increased the workload of an already understaffed authority. Such 'Big Science' contexts thus contrasted starkly with the age-old arrangement of scientific talent into organizations like universities or the Royal Society, which allowed experts a great deal of moral and intellectual space, albeit on a comparatively meagre budget.

Somewhat controversially, Cherwell's biographer, in a view supported by Gowing, has claimed that the Atomic Energy Authority was 'won single-handed' by Cherwell and must therefore be regarded as his 'political monument'.[65] Such assertions are not completely accurate, as Cockcroft also supported the move and influenced his wavering staff to do likewise, but it is nonetheless difficult to deny that Cherwell's unique position inside political circles greatly catalysed change.[66] The formation of the UKAEA therefore reflected not merely an organizational foible but was part of a significant trend in which national governments began to concede that departmental organization was inadequate when priority fields required ever-greater funding and substantial institutions. Accordingly, men of technical education were increasingly required to oversee such establishments, reducing government to offering policy directives. As Sir Frederick Brundrett, chief scientific adviser to the Ministry of Defence, warned government: 'unless you incorporate the scientist as a full member of your team you are not turning out your First Eleven, and the mistakes that are made will be more numerous and have more serious consequences'.[67]

International Role

A final notable feature of the scientist-diplomats was their effectiveness in conducting scientific relations across national borders. Cockcroft, Chadwick and Blackett assumed an active role in promoting international co-operation in the wider atomic field, notably during the 1953 negotiations over CERN, the European fundamental physics laboratory project. As government departments prevaricated over the decision to participate, the scientist-diplomats used their influence to overcome political uncertainty and eventually emerged successful.[68] Furthermore, Cockcroft undertook lengthy lecture tours, promoting atomic energy and discussing the progress made at Harwell with audiences as far away as New Zealand and Australia. As the decade drew on, he also visited states behind the Iron Curtain, where he was consulted on technical matters and asked about possible collaboration with the UK.[69] Thus, whilst obviously restricted by security considerations, these visits nonetheless demonstrated the ambassadorial skills of individual experts and their ability to represent an entire scientific sub-field.

In any case, by the mid-1950s Britain's atomic industry had become world-leading, with a White Paper in February 1955 identifying a role for civil stations both in the domestic energy mix and as an exportable commodity.[70] As John Krige has highlighted, British atomic goods also stole the show at the Geneva science conference that summer, while Harwell arranged for scientists from thirty-one countries to participate in a day-trip to the AERE in order to meet eminent British researchers and inspect engineering equipment.[71] In pioneering a new industry of global standing, Britain's atomic engineers consequently ensured that they would require frameworks to deal with exports and other foreign policy concerns. Thus, the origin of the scientist-diplomat as an agent of international diplomacy is inextricably bound to his position as an eminent adviser domestically, returning us to the original definition of the scientist-diplomat outlined in the introduction to this chapter.

Nevertheless, despite their positioning within new organizations, the influence of scientist-diplomats was not unlimited. The contemporary trend towards European political and economic integration resulted in the Treaty of Rome in 1957, placing Britain in a conundrum as Whitehall slowly warmed to the idea of participation. Such notions contradicted the atomic scientists' scepticism of Euratom, the Atomic Energy Authority established alongside the EEC, and UKAEA officials were keen to preserve their hard-won technological lead. Participation in an integrated authority threatened the possibility that Britain should divulge its research to other parties, and Plowden therefore recommended that Britain's atomic interests were better-served by bilateral agreements instead of supranational organizations.[72] However, in 1962, Prime Minister Harold Macmillan was persuaded by the wider attraction of Europe to override his domestic authority, marking a logical end point for this analysis by highlighting atomic energy's depreciating value as a commodity. Despite re-organizing atomic energy into a corporational structure, politicians now felt justification in overriding scientific advice in pursuit of greater goals, and what was once the shining jewel in Britain's technological crown was relegated below wider-reaching macroeconomic concerns.

Conclusion

To conclude, this chapter has traced the development of a unique hybrid character who straddled the divide between the scientific and political worlds during the mid-twentieth century. In a period of general upward mobility for scientific representation, atomic energy, as a priority field, catalysed fundamental changes in the negotiated power-balance between science and state. Consequently, to operate successfully in a new environment where governments incorporated scientific assistance, expert elites evolved into the 'scientist-diplomat', the man as capable in the laboratory as the committee room.

The particularity of atomic energy in this regard can be identified in the ripe conditions that provided the stimulus for change. Britain developed the technology, and subsequently such science-government relationships, earlier than other states because the political environment demanding atomic weapons coincided with economic conditions requiring cost-effective electricity. This was a rare confluence not seen elsewhere so soon after the war; the United States for instance, had no immediate need of civil atomic energy whilst smaller European nations, notably the Netherlands and Norway, desired economical power without weapons. However, although civil atomic energy was politically popular for presenting a novel solution to fossil fuel shortages, its development came at the cost of introducing a technology where the average politician could scarcely understand the method, merely the result, thereby compelling scientists to assume greater influence. Consequently, the development of this new expert breed highlights how the scientific plane adapted to its political restraints rather than vice-versa: we identify the 'scientist-diplomat' rather than the 'politician-scientist'. The select band of elite physicists who could advance atomic energy in the United Kingdom therefore assumed substantial importance as they developed a sensitive field.

Accordingly, the main focus of this chapter has been to identify the administrative and political nature of the expert role of the atomic scientist, and thereafter to trace the evolution of expertise through changing professional structures. Concentrated by wartime projects, atomic scientists refined a collective conscience, in turn stimulating scientific associations to effectively represent scientific views at both public and official levels. Engendered with a substantial moral credit by the contentious nature of his work, the scientist-diplomat was consequently empowered to better represent scientific ethics, placing him occasionally in opposition to government policies. Against this backdrop of scientific consciousness, eminent scientist-diplomats were introduced to the traditional political environment of the government's committee system and expected to offer advice alongside civil servants and ministers representing diverse interests. Such increased exposure to individuals and processes in the governmental world therefore provided political education to those elites invited to represent their field, while simultaneously constructing a forum to discuss concerns. Importantly, the committee system distilled the interface between politicians and experts into its purest form, enabling the relationship to attain equilibrium where experts exercised a moderating (or occasionally reforming) influence on government initiatives.

Nonetheless, the construction of nationally funded research establishments mandated that scientific influence remain somewhat restricted, and the subjugation of Harwell, Risley and later Aldermaston to direct state control clearly demonstrated Whitehall's determination to see Britain's atomic objectives kept under close supervision. Despite being granted considerable freedom of research, the scientist-diplomats also became restricted by departmental inertia, reducing their effectiveness as policymakers. Thus, although an important feature of any democratic state's operations regarding science, as Dominique Pestre contended, the political contestation of scientific opinion was struggling to locate a mechanism acceptable to both scientific and lay groups.[73] Change would require a political spearhead, and this was provided by Cherwell, who delineated a unique sub-group within scientific diplomacy via his preference for political office and direct communication with the Cabinet. Once the Conservatives regained power in 1951, Cherwell was able to deploy his *personal* and proximal influence (as opposed to the institutional or representative power obtained by other scientist-diplomats) with Churchill to affect change. However, despite mobilizing the prime minister to support greater independence within atomic organization, Cherwell was never elected to public office, highlighting the undemocratic trends inherent in this technocratic model, and further demonstrating the difficulty of incorporating high-performance expert advice into British policymaking.

The product of many years' negotiation, the formation of the UKAEA in 1954 consequently marked a clear attempt to increase the national atomic 'effort', both in funding and efficiency, by reconfiguring atomic organization at the interface of state, science and industry. The 'troika', as Hinton called it, of research, industry and weapons were combined and coordinated by a board which represented each aspect of the technology to a mediating chairman of considerable experience.[74] As such, the scientist-diplomat became a key administrative and political figure, gaining greater control over internal policy whilst accepting overall budgetary surveillance from government. An atomic energy executive was also established to debate opinion between the troika, increasing the quality of expert advice. The Authority therefore adopted a curious hybrid of several organizational traditions including academic, governmental, industrial and civil service, helping it to thrive despite an unsteady intake of manpower. In this way, the scientist-diplomats were arranged into a significant structure whose longevity helped them to better advise governments which were now obliged to take decisions that ran decades beyond their democratic mandate.

Thus, the rise of the scientist-diplomat ensured that in Britain, the expert was now a permanent fixture opposite government. By 1964, the ruling Labour Party had appointed a chief scientific adviser to the government itself, alongside a full Ministry of Technology. Under this modernist umbrella, Prime Minister Harold Wilson famously promised that he would build a new, technocratic Britain forged in the 'white heat of revolution' where there would be 'no place for restrictive practices or outdated methods on either side of industry'.[75] What he failed to mention was that, in atomic energy at least, Britain was already well ahead of the trend.

8 THE REFORM TECHNOCRATS: STRATEGISTS OF THE SWEDISH WELFARE STATE, 1930–60

Per Lundin and Niklas Stenlås

The history of modern Sweden has centred around the creation of the welfare state. It has been argued that a strong interventionist state, a large public sector and the extent of its welfare commitment has set Sweden apart from other European countries. Those traits have also attracted interest from many foreign observers who have coined the notion of a Swedish or Scandinavian model.[1]

By emphasising the role of a new group of 'strategists' in the planning and creation of the strong state during and after the Second World War, we attempt to add a new dimension to the interpretation of the Swedish welfare state's emergence. The role of engineers, planners, scientists and other professionals has generally been thought of as secondary to that of political leaders and visionaries. Experts may have been important, but seldom in other capacities than as executors of political directives.[2] We do not believe that this is an accurate description.

The many state initiatives in society, science and the economy are well known from earlier research within each sector, but when taken together, a new pattern emerges.[3] It is possible to identify the contours of a new state commitment that emerged during the latter part of the interwar period and was institutionalized during the following decades. Furthermore, the development within each sector demonstrates a number of similar features: a rather small group of individuals – typically architects, economists, engineers, planners or scientists – lobbied the government to set up government committees on which they served as experts. Once on the committees, they led the planning and surveying of the needs, and proposed the founding of institutions of which they were later appointed leaders and, thus, became the implementers of their own plans and visions. In addition, they often had academic careers, thus gaining a scientific legitimacy for the reforms they carried out as committee members or institute directors. They mainly defined themselves as apolitical and disinterested experts, employing scientific methods to overcome political and social conflicts of interest. For them, science became a means to do politics.

If the course of events within all these sectors is studied simultaneously, the same individuals in many cases assume central roles within different sectors.[4] Most of them had a similar educational background. They were young academics, typically architects or engineers from the Royal Institute of Technology in Stockholm, economists from the Stockholm School of Economics, physicians from the Karolinska Institute in Stockholm, social scientists from Stockholm University College or scientists from Uppsala University. Some were active Liberals or Social Democrats, but most were professionals first, and politicians second, if at all. Often they had become acquainted at the university, where they sometimes formed informal networks. Their common beliefs, norms and values were strengthened by their membership of professional organizations, some of them with close links to business and industry, such as the Royal Swedish Academy of Engineering Sciences, the Swedish Association of Engineers and Architects, the Swedish Society for Town and Country Planning and the Swedish Road Association. They shared an 'engineering mentality' and the conviction that it could be applied everywhere in society. The American historian Charles S. Maier describes this professional creed as an ideology of productivity originating in America but adopted in most of Europe from the First World War on. It was necessitated by the war efforts in both world wars as well as by the need to reform industry and the economy in the wake of the Great Depression. As Maier puts it: 'The engineers would be summoned to impose optimality upon society as they did in the factory.'[5] They believed in the decisive role of the state, maintaining that the state should both take the initiative and hold a central role in the development of society. This faith in the state was above all expressed in a belief in planning.[6] Thus, in this respect, they can all be called planners.

In this chapter we aim to investigate a group of architects, economists, engineers, planners and scientists who populated the government committees, headed the institutions, advised the government and eventually advanced to director generals or state secretaries in the new, institutionalized 'scientific' state. We argue that they were not only an instrument of the strong, active state, but often also the primary actors, initiators and architects of the many state initiatives, and we refer to them as reform technocrats.

This chapter examines three closely intertwined questions that we regard as central to explaining and understanding the reform technocrats' role in the establishment and institutionalization of the post-war state initiative. First, we describe and analyse the reform technocrats' identity. Who were the central actors? How were they related to each other? To what extent did they have a common educational background? What career paths did they take? Which positions and institutions were considered key? How were people recruited to these positions? Second, we examine closer the reform technocrats' strategies and actions. What courses of action were taken and strategies used to influence politics? At this point, it is important to emphasize that the reform technocrats were part of a specific historical context which framed their activities. This context consists of 'structural' factors, including changing economic conditions, Social Democratic hegemony, the particular structure of the Swedish state (small ministries, independent government agencies and independent government committees), corporatism and the Cold War. Although the reform technocrats were instrumental in initiating and shaping many of the post-war reforms, their influence was far from absolute, varying over time and also from sector to sector. Third, we therefore empirically investigate the borderland between politics and the economy, i.e. the boundaries of the reform technocrats' legitimate domains. How far did these go? And how did they change over time? Which strategies did the reform technocrats employ to expand their legitimate domains?

To sum up, it is important not only to describe and analyse the deeds of the reform technocrats in order to understand the establishment, expansion and consolidation of the strong post-war state, but also to demonstrate the claim that the post-war reforms were not solely prompted by the political parties' ideologies and activities.

Welfare State Reforms and the Problem of Agency

Typically, the welfare state project has been seen as a natural outcome of the ideologies of Sweden's social movements and political parties, especially with respect to the Labour movement and the Social Democratic Party.[7] However, there are variants of this theme where the support for the welfare project by the non-socialist parties and even the employers has been stressed.[8] There is also a structuralist strain in the historiography of the 'Swedish model', where the 1930s class compromise between labour and capital has been regarded as explaining the rise of the Swedish welfare model.[9] It is worth mentioning that it is not first and foremost historians, but scholars from other disciplines who have formulated the grand theories of the emergence of the Swedish welfare state.[10]

Swedish historians, on the other hand, have often focused on individual political actors among whom the 'social engineers' have played a prominent role as visionaries, reformers and pioneers.[11] However, it is difficult to see that the scholarship on social engineers has had a significant impact on the historiography of the Swedish welfare state. Perhaps this is due to the fact that the social engineers identified by scholars are so few that they can be counted on the fingers of one hand. Besides the iconic intellectual couple, Alva and Gunnar Myrdal, scholars have found it difficult to pinpoint other social engineers or welfare state experts.[12] Generally, this scholarship discusses the 1930s and the radical ideas flourishing by then. It rarely contextualizes the social engineers;

rather, it treats them – in the words of Karl Mannheim – as a kind of free-floating intelligentsia (*Freischwebende Intelligenz*).[13] A more fitting description is perhaps 'public intellectuals', a term originally used to describe a new group of intellectuals who entered the US public debate and sector around the time of the First World War.[14] But as the Norwegian sociologist and historian Rune Slagstad has pointed out, these radicals and their ideas did not really resonate with the state until the new political order was firmly in place. In other words, they needed institutions and legislation so that reforms could be realized.[15] Although some of the actors we are interested in can be depicted as social engineers or public intellectuals, we are primarily interested in them as initiators and leaders of institutions and later of specific reforms.

The reform technocrats described above are therefore a significantly broader category than the social engineers and the public intellectuals,[16] who typically had a social scientific background, were explicitly politically and ideologically committed and sought to introduce social reforms. On the other hand, the reform technocrats were seemingly apolitical professionals possessing, or claiming to possess, administrative, technological and/or scientific expertise, and who sought to overcome political or social conflicts of interests by means of scientific methods. This is also, we believe, the reason that the former, unlike the latter, has been visible and of interest to authors writing the history of twentieth-century Sweden.

This chapter gives three examples that illustrate the mechanisms of reform technocracy, namely housing and planning reforms, higher education and research reforms, and primary and lower secondary education reforms. Although the examples in their own respects are well known from earlier research, the interesting point is, we contend, the pattern they reveal.

Government Committees, State Agency and Social Reforms

The specific structure of the Swedish state (small ministries, independent government agencies and independent government committees) largely determined the reform technocrats' line of action. The ministries had very little in-house expertise, and therefore delegated the task of charting the course of state action in different economic and social sectors to agencies and committees. The government appointed the director general of each agency, but otherwise rarely interfered in the agencies' affairs. Consequently, the agencies established themselves as powerful and autonomous actors during the period studied. Normally, parliamentarians from different political parties, representatives of corporatist organizations and independent experts worked on the committees, which were not only 'parliamentary' but also 'expert'. During the post-war decades, the number of experts involved rose dramatically.[17] Usually, a large number of bodies, such as political parties, corporatist organizations and government agencies reviewed the work of the committees. This process not only helped shape the intended reforms but also served to firmly establish them in Swedish society. The Swedish committee system has been acknowledged as having a key role in influencing government policies and social reforms.[18] The committees often sat for several years, in some cases even decades, with many providing important intellectual platforms for the emergence of social sciences in Sweden.[19] This state structure gave ample room for government agencies and committees to exert an influence using their expertise.

Several government committees on housing, land use, town and country planning, and social reforms were appointed during the 1930s and 1940s. The reform technocrats not only controlled them but also were often the driving forces behind setting them up. When the Housing Committee started its work in 1933, its appointment had been due to an energetic campaign undertaken by the economist Gunnar Myrdal and the architect Uno Åhrén, whose close connections with leading Social Democrats helped. Myrdal had joined the Social Democratic Party in 1932, the very same year as the party came to power. Myrdal later recalled how the minister for health and social affairs, Gustav Möller, gave them 'carte blanche to gather like-minded friends to form an official committee and involve other friends as a secretary and experts'.[20] And so they did. One 'friend' was the economist Alf Johansson, the committee's principal secretary and later its chairman. In addition, the master builder Olle Engkvist, the architect and managing director of the Tenants' Savings and Building Society Sven Wallander, as well as the city architect of Stockholm Sigurd Westholm, all joined the committee. However, Johansson was the committee's undisputed leading figure. The committee sat for fourteen years, and with its two final reports it set the tone for the post-war housing policy. Per Nyström, state secretary at the Ministry of Health and Social Affairs and an old friend of Johansson, oversaw the work on the subsequent government bill.[21] Following the committee's proposal, the Royal Board of Housing was established in 1948, with Johansson as its first director general, a position he held until 1960. As director general of the housing sector's most powerful agency, he could guide and oversee the proposed reforms.[22]

As the example suggests, the findings of the committees were, in many cases, decisive for shaping government policies and social reforms within the different sectors in question. Often a set of government committees investigated closely linked issues. For instance, committees on land use, city planning, municipal and regional policy reforms and higher education in planning worked more or less in parallel with the Housing Committee.[23] It was not uncommon to find the very same individuals sitting on several committees. Uno Åhrén held such a key position in the aforementioned group of committees. They also recommended or picked each other when expertise was needed for similar or newly appointed committees. Thus, many of Åhrén's former students (he taught town and country planning at the Royal Institute of Technology) and younger colleagues (he was head of the Town and Country Planning Office in Gothenburg, and later managing director of Riksbyggen, a building company owned by the building unions, local housing associations and other national cooperative associations) replaced them on the next generation of housing, land use and planning committees.[24] In some cases, the networks were even shaped by matrimonial ties. Brita Åkerman, who worked on several committees and headed the Home Research Institute, and Alf Johansson were a married couple, as were the social reformists, Alva and Gunnar Myrdal.[25] Quite a number of them held academic positions. Johansson taught at Stockholm University College, where he eventually became professor of economics. Gunnar Myrdal held since the early 1930s a chair at the same university. And Åhrén became Sweden's first professor of town and country planning at the Royal Institute of Technology. Such positions increased their scientific credibility and justified their involvement in social reforms.

The committees' findings often resulted in institutional reforms, such as the creation of new institutions or laws, or the strengthening of existing ones. Newly created agencies like the Royal Board of Housing or reformed agencies like the Royal Board of Building achieved influential positions. The former, for instance, was foremost an agency that managed state loans for housing construction. Since about 65 per cent of Sweden's four million dwellings were constructed during the post-war period, and over 90 per cent of the dwellings built between 1961 and 1975 were financed by state loans,[26] the new terms of lending as formulated by Johansson and his peers had a decisive influence on the post-war construction boom.

The example of Johansson illustrates how this group of individuals rather often achieved leading positions in the institutional framework they had shaped. As we shall see, this pattern is repeated in other sectors as well.[27] A demand for expertise was created, taking various forms, such as in-house expertise or consulting. The architects of the reforms frequently hired each other or their former students for such services. A clear-cut example is the Building and Planning Act of 1947, which allowed municipalities to control land use with the help of master plans. The Royal Board of Building, the authority responsible for planning commissions, provided instructions on how contractors should interpret the Act in order to obtain building permits. The Act was the result of the 1942 Committee on City Planning, a key member of which was the architect Sune Lindström, a younger colleague of Åhrén. After serving on the committee, he worked for the major Swedish consulting company Vattenbyggnadsbyrån (VBB), where he was responsible for developing master plans in accordance with the new act. The VBB was frequently hired by municipalities lacking the necessary expertise.[28]

Although the above-mentioned reform technocrats had good contacts with the Social Democrats, and sometimes even considered themselves the (political) avant-garde, Conservatives and industrialists very often supported their ambitions. For example, the managing director of Skånska Cementgjuteriet (later Skanska), one of Sweden's largest building companies, Ernst Wehtje, championed the Housing Committee's, in many aspects, radical reforms. Besides being an industrialist, Wehtje was a Conservative member of the Swedish Parliament. He was also the former chairman of the Federation of Swedish Industries as well as of the Royal Institute of Technology.[29] Not all social reformists had this kind of support, but in other sectors, such links with industrialists were the rule rather than the exception.

Interest Organizations, the Active State and the Shaping of Cold War Research

The breakthrough in creating a new, socially oriented science policy saw corporative and professional organizations like the Royal Swedish Academy of Engineering Sciences, the Swedish Association of Engineers and Architects and the Federation of Swedish Industries lead the way. The representatives of the organizations – most notably the electrical engineer Edy Velander from the Royal Swedish Academy of Engineering Sciences and the civil engineer and industrialist Sten Westerberg from the Association of Engineers and Architects – lobbied the government for reforms, arguing that engineering science research had a key role in industrial development and that the state and industry were mutually responsible for promoting technological research. As a result, the Malm Committee, named after its forceful chair, the civil engineer and director general of the Royal Waterfall Board Gösta Malm , was established in 1940. Malm was a former managing director of Skånska Cementgjuteriet, and twice served as a non-party government minister. The committee's aim was to propose new ways of organizing and financing science and technology research. The lobbyists themselves drafted the committee's terms of reference and appointed its members. Consequently, both Velander and Westerberg sat on the committee together with leading scientists, such as the 1926 Nobel Laureate in chemistry, Theodor (The) Svedberg.[30]

The Malm Committee worked swiftly, producing three complete government reports within eighteen months. The government adopted the proposals put forward, and, as a result, the Swedish Council for Technical Research and the Swedish Committee for Building Research were founded in 1942.[31] The establishment of these two bodies proved paradigmatic for the organization of government-sponsored research in Sweden. National research councils for medicine, natural sciences, social sciences, agricultural technology and traffic safety followed. Three of the eight board members of the Council for Technical Research were selected from the Malm Committee, namely Velander, Westerberg and Svedberg. Thus, the very same individuals who had originally proposed the council and lobbied for its establishment were appointed to head it. In particular Velander, – who had become managing director of the Royal Swedish Academy of Engineering Sciences in 1941 – used this influential position for establishing new research institutes and promoting state-industry alliances in such different areas as computing technology and food technology.[32]

Also military research was reformed during the war years, with the Royal Swedish Academy of Engineering Sciences and Velander's networks again playing a key role. Only a few weeks before the outbreak of the Second World War, the National Commission for Economic Defence, a government body founded in the late 1920s to plan and ensure the workings of the Swedish economy in the event of war, blockades and so forth, had asked the academy to compile an inventory of national science resources, in particular physics and chemistry. The academy commissioned the physicist Rolf Sievert, who in turn enrolled his close friend Manne Siegbahn, the 1924 Nobel Laureate in physics and one of the Swedish scientific community's central figures. Together they investigated how physicists could help strengthen the country's defence. Sievert and Siegbahn managed to convince Minister for Defence Per Edvin Sköld of the strategic importance of physics research, and their efforts led to the foundation of the Institute of Military Physics in 1941.[33] Two years later the government set up a provisional body, the Defence Research Board, as part of its wartime administration. The board's aim was to coordinate the existing research efforts and to devise a future overarching research organization for the entire defence sector. Siegbahn, who had been appointed a board member of the Council for Technical Research in 1942, was selected as its chairman, and Sievert as one of its members. The board's work led to the defence research bodies, among them the Institute of Military Physics, merging to form the Swedish Defence Research Establishment in 1945.[34]

The Swedish Defence Research Establishment became – together with the formation of a modern arms industry, with the aircraft industries as the technically most important part – the most lasting effect of the wartime armament, growing to become by far the largest research organization in Sweden during the 1950s and 1960s. The Swedish nuclear weapons research programme, involving almost a quarter of the agency's workforce in the mid-1950s, played a key role in this expansion.[35]

Håkan Sterky, the first chairman of the Swedish Defence Research Establishment, epitomizes the close ties between the agency and the nuclear weapons programme. Sterky was an electrical engineer from the Royal Institute of Technology, where he eventually came to hold a chair. Furthermore, he was director general of the large Swedish Telecommunications Administration for more than two decades. Sterky sat on the 1945 Atomic Committee (as did half of the agency's board), where he, together with the physicist and Prime Minister Tage Erlander's informal science adviser Torsten Gustafson, and other prominent scientists, such as Manne Siegbahn, The Svedberg and Arne Tiselius (the 1948 Nobel Laureate in chemistry), investigated the possibility of a Swedish nuclear energy programme. The committee's proposal led to the foundation of the state-owned company, AB Atomenergi. Along with several members of the committee, Sterky sat on the company board, a position he held for twenty-two years.[36] Science policy was an area where the reform technocrats exercised an almost-unparalleled influence vis-à-vis political actors. In other areas their legitimate domains were more constrained and their claims contested. A case in point is primary and lower secondary education.

The Entanglement of Professional and Political Agency

In the early 1930s, Sweden had a parallel school system. It prepared only a minority of the children for higher education, and the majority for work at the age of 14. The system irritated many Social Democrats and Liberals, and particularly within the Social Democratic Party there were visionaries striving to carry out educational reforms. Alva and Gunnar Myrdal, for instance, expressed a desire to form new citizens for the new society they envisioned. They saw education as a central tool for social reform. The majority of the Social Democratic Party were not that interested in school reforms, however. Quite significantly, Arthur Engberg, the Social Democratic minister for education, did not endorse any major educational reform.[37] In fact, no attempt whatsoever was made during the 1930s to reform the school system.[38] In retrospect, this has puzzled historians, who have seen the dismantling of the old parallel school system as a central component of the construction of the welfare state and social democratic ideology as the major force behind the reform.[39]

During the Second World War, Sweden had a coalition government, with the Conservative Gösta Bagge the minister for education. In 1940, he appointed a government committee to chart the needs of primary education and propose a root-and-branch reform. Bagge served as chairman of the committee – an arrangement that signalled the government's commitment to the issue. The 1940 School Committee was a typical expert committee. The majority of its fifteen members had a professional background in primary or lower secondary education, and many were well known, as a result of having participated in the contemporary school debate.[40] Between 1944 and 1947, the committee published no fewer than twenty volumes (appendices included) – a total of 4,092 pages. The committee recommended a unitary school system with an eight-year compulsory schooling, but it could not agree on the critical issue of when the children aiming for higher education should be separated from those who should be prepared for work.

On 31 July 1945, almost three months after the end of the Second World War, the coalition government was replaced by a Social Democratic one. Tage Erlander, former state secretary at the Ministry of Health and Social Affairs, was appointed minister for education, a field in which he was, however, not well versed, and he therefore chose Josef Weijne as his state secretary. Weijne had been a Social Democratic member of the Swedish Parliament since the mid-1920s and the party's foremost expert on primary and lower secondary education. Between 1934 and 1945, he had served on no fewer than thirteen government committees, most dealing with educational policy, and had chaired four of them. In

1939, Weijne, a schoolteacher by profession, advanced to head of division at the Swedish National Board of Education. Thus, even if Weijne was an active Social Democrat, he was first and foremost an expert.[41] Weijne participated in an internal working group set up by the Social Democratic Party in early 1945; its aim was to formulate a Social Democratic education policy before the expected resignation of the coalition government. Weijne and the working group argued for a general school system for all children. Education should be democratized, and social, economic and geographical differences erased. The ideas they presented were not included in the Social Democratic Party programme adopted later the same year, however. But when appointed state secretary, Weijne was able to realize the ideas the party had been uninterested in including in its programme.[42]

Weijne and Erlander worked very well together. 'It really feels extremely reassuring to have Weijne by my side', Erlander wrote in his diary, 'and I am happy that he also seems comfortable with me'. Weijne quickly became the strategist for primary and lower secondary education. He convinced Erlander that it was necessary to replace the 1940 School Committee – which was still sitting – with a new one as soon as possible.[43] On New Year's Eve of 1945, Erlander had Weijne and Per Nyström, state secretary at the Ministry of Health and Social Affairs, and their respective families over for dinner. 'We came to the conclusion', Erlander wrote, 'that the findings of the Bagge Committee [the 1940 School Committee] should hardly be circulated until we had a firmer grasp of our own policy'.[44] Early the following year the government set up the 1946 School Committee, whose very appointment while the old one was still working with primarily the same tasks, is probably unique in the history of Swedish government committees.

Just like the preceding committee, the minister for education, in this case Erlander, appointed himself chairman, thereby emphasizing the task's high priority. The vice chairman was Weijne. This time, however, it was a parliamentary committee and all except one of the remaining members were politicians. Even if politically appointed, the members were nonetheless considerably knowledgeable about education issues. With the exception of Erlander, all of them had professional experience of the education field. No fewer than eighty experts were attached to the committee.

Weijne was the new committee's prime mover, convincing Erlander that a committee needed to be appointed. Moreover, the committee's terms of reference were directly based on the draft education policy he had written for the Social Democratic Party's internal working group some years earlier but which had never been adopted in the party programme. Weijne also influenced the selection of the committee's Social Democratic members. Of the four Social Democratic parliamentarians, two had, together with Weijne, participated in the party's internal working group. It can thus be argued that they were already committed to the policy outlined in the committee's terms of reference.[45]

When Erlander in October 1946, after only nine months as minister for education, replaced Prime Minister Per Albin Hansson – who had unexpectedly died – Weijne succeeded Erlander as both chairman of the 1946 School Committee and minister for education. Thus, as chairman, Weijne would deliver the committee's government report in 1948, and as minister he would oversee that its proposals would become a government bill. The physicist Ragnar Lundblad wrote the government bill, an impressive piece of over 590 pages. Lundblad had a leading position at the Swedish National Board of Education, and had been engaged as an expert sitting on both the 1940 and 1946 School Committees. In 1950, Weijne presented the bill to the government and, later, defended it in the Parliament.[46]

The workings of the 1946 School Committee, its government report and the subsequent bill outlined the main features of primary and lower secondary education up until the late 1980s, if not longer. Weijne had already died in 1951, but the educational politicians and reform technocrats who succeeded him, and had mostly gained their experience under his chairmanship of the 1946 School Committee, would eventually implement the proposed reforms, the most important being the establishment of a nine-year integrated school in 1962.[47]

This final example differs somewhat from the previous ones, in that the protagonist, Weijne, was a Social Democrat with a central role within his party. His political position was not strong enough to affect the party's policy, however. But when he was suddenly appointed state secretary, he was able to bypass the party and shape government policy by other means. As the minister's right hand, he could influence the selection of the committee members, the agenda of the committee, as well as which experts were chosen to furnish the committee with expertise. Thus, his course of action resembled more that of a reform technocrat than that of a politician.

Conclusion

In the examples, we have identified key actors that had expert status and could access the political system and affect legislation and reform policy. They did not do so individually, but together with other like-minded friends, colleagues or allies. What united them was a belief in the necessity of an active state for reforming society. Moreover, in the process of introducing knowledge-based reforms, they also paved the way for their own careers. We have called them reform technocrats.

In all three cases, the reform technocrats' participation in government committees was decisive for their ability to shape reforms. While on the committees, they used their and their colleagues' expert status to influence the committees' work and findings. In all three examples, even the initiative to appoint the committees came from one or several of the actors we have identified as reform technocrats. That they made a public administration career in parallel with 'their' reform is also important because it allowed them to follow the reform and oversee its implementation.

The ability to influence leading politicians to appoint committees or to accept their findings has been an important feature of all the reform technocrats studied. Without this ability, their impact on politics would be down to chance rather than strategy. How leading politicians were influenced varies from case to case, however. With regard to housing policy, a group of reform technocrats worked closely together and persuaded the minister for health and social affairs to appoint a committee on which they subsequently worked, thus enabling them to formulate the terms of reference and influence the appointment of experts. In the case of primary and lower secondary education, on the other hand, a professional teacher, who had been a Social Democratic member of the Swedish Parliament and with long experience of public administration and committee work, was appointed state secretary at the Ministry of Education under the leadership of a new and inexperienced minister. The state secretary, thus, combined the roles of an expert, a strategist and a political connection. When it came to science policy, government policies were susceptible to lobbyism from reform technocrats with links to interest groups; they not only influenced the appointment of a committee but also wrote its terms of reference and secured seats on it. Eventually, when institutions were created based on the committee's recommendations, the lobbyists themselves were appointed to leadership positions within these institutions.

The system of government committees occupied a central role in Sweden's political decision making, providing the reform technocrats with an 'institutional space' through which they could access, influence and shape government policy. This space was open to negotiation and accessible to different actors, such as the Cabinet, interest organizations, political parties and experts, with the Cabinet and the reform technocrats being the most influential. The Cabinet had the power to appoint committees and select their members. In addition, it wrote the government bills and put them before the Parliament. The power of the reform technocrats was based on their claims to possess scientific legitimacy and organizational fiat. This was further strengthened by their positions within academia and public administration as well as within interest organizations and political parties. Moreover, their careers enabled them to follow a reform from its inception to its realization. In fact, their careers were often, albeit not exclusively, closely linked with the reform in question. This distinguished them from other actors linked to the committee system. Neither political parties nor interest organizations had that opportunity or personal incentive.

The question of reform technocracy's legitimate domain is somewhat difficult to answer. It is important to note that merely being an expert or even having access to a committee is not enough for an effective reform technocracy. There must also be a political mandate to act. This has to be secured through political alliances. However, it seems that the dividing lines between technocratic and political actors are blurred and may shift from one sector to another. But all three examples demonstrate that the reform technocrats were considerably more well informed and strategic in their actions than their political allies.

9 RATIONALIZATION COMES TO ROME: EXPERTISE IN LABOUR MANAGEMENT AT THE THIRD INTERNATIONAL CONGRESS, 1927

Jennifer Karns Alexander

> But the final note resounding in our ears was a clarion call to earnest, devoted, consecrated work, directed and enlightened by the dictates of the uncompromising Spirit of Science.
>
> Eleanor Bushnell Cooke[1]

An imperial call of trumpets announced the arrival of Benito Mussolini at Rome's famous Campidoglio in the autumn of 1927, where he addressed the concluding ceremony of the Third International Congress on the Scientific Management of Labour. Soldiers in bright uniforms lined the stairways, and more than 2,000 people filled the Senate Chamber: diplomats, government officials, senior military officers, conference delegates and their guests. Eleanor Bushnell Cooke described Mussolini's speech: he 'left no one in doubt as to the value he placed on Scientific Management, nor of his sympathy with the Congress just being brought to a close'.[2]

Cooke, reacting to Mussolini's address, described the spirit of scientific management as uncompromising, and offered for its mode of action not persuasion, but something stronger: science dictates, she wrote. This phrase, science 'dictates', encapsulates the point this paper will make about performances of expertise at the Rome Congress: that such performances – debates on grand stages before audiences of thousands, hurried presentations in small lecture halls, mimeographed papers circulated in hard-copy but not presented – were not negotiations through which scientific expertise was constructed, but were, instead, demonstrations of the successful objectification of the working people upon which experts based their claims to expert knowledge. Performances in Rome displayed not the construction of expertise, but its function. Working people and their labour became variables, manipulated by management officials and consultants in demonstrating the effectiveness of their specialized knowledge. Such objectification was not incidental but definitive in the performance of management expertise.

This article responds to several limitations of current scholarly understandings of expertise as performance. It analyzes a congress sponsored by a fascist patron, Mussolini, who enthusiastically endorsed the principles it celebrated; it thus stands in contrast to a prevailing scholarly focus on expertise within liberal-democratic societies.[3] It also stands in contrast to descriptions of expertise as developing from negotiated or mediated performances.[4] Useful discussions of negotiation and mediation assume a negotiating partner with at least some effective agency, yet precisely what the Rome Congress celebrated was the demise of the effective agency of a potentially provocative audience of working people.

What follows is a close reading of the Rome Congress, and of the details of the presentations made by leaders in labour management; it is a close reading of their performances of expertise. This reading highlights the effectiveness of these performances, by looking not at the particular social configurations that gave rise to each conference contribution, or even at the particular context of the congress itself, beyond establishing that it was a significant event in the developing specialized field of labour management. Such a close reading stands in contrast to both historical and social science approaches to expertise by seeing its meaning not in the particular social, economic or political contexts or networks within which it was performed, but in the content of the performances themselves. Anchoring expertise in the particular constellations of factors within which it occurred is to anchor any analysis of it to the same constellations; when they no longer obtain, an analysis so anchored is no longer valid. A close reading yields an analysis that can cross social, economic and political boundaries, and treats performances of expertise as simultaneously performances of content.

In Rome, that content was disciplinary. Expert knowledge in labour management was not at its root a matter of negotiation and trust, but of ensuring required behaviours, and expertise lay in obscuring the unique features that characterized individual working persons, and thus objectifying and operationalizing them. Objectified knowledge of labour operated, materially, both to display and to enhance its own effectiveness. The key to expertise at the Rome Congress was performances displaying techniques that disciplined working people, by ensuring that they behaved as desired, while simultaneously measuring how effective such discipline was.[5]

Expertise, then, is more than specialized knowledge or skill. Expertise operates at the intersection of specialized knowledge and skills and deeply-held ideals of how the world should work. The performance of expertise is its embodiment in actual, material engagement with the world. In the case of labour management at Rome, performed, embodied expertise helped to reveal the ideal of the management expert as an active authority, and workers as objects of expert manipulation. It may be that the expertise embodied at the Rome conference was a mature form, in which elements of authority had congealed and moved beyond earlier stages at which it was open to influences from critical audiences; I will conclude by suggesting that scholars consider not only how expertise is constructed, but how it becomes hardened and increasingly impervious to critical influence.

Rome, 1927: The Third International Congress on the Scientific Management of Labour

The Rome Congress was a significant event in the developing field of labour management. Two previous conferences had been held, in Prague (1924) and Brussels (1926), and attending them was an honour. The congress brought together two strands of a movement less than a generation old. From the Americas came experts in scientific management, who emphasized personal efficiencies and efficiencies at the level of the firm; from elsewhere came proponents of the rationalization movement, which offered a broader vision of efficient and orderly society. European planners emerging from the First World War envisioned rationalization as a way to eliminate scarcity by eliminating waste and by re-organising production and distribution to help ease civil unrest.[6] Rationalization was more labour-intensive and less capital-intensive than scientific management, and its advocates were often suspicious of mass production. Scientific management was most famously associated with the mechanical engineer Frederick Winslow Taylor, rationalization with industrialists such as Carl Friedrich von Siemens. The practical effect of both movements is unclear. Both influenced industrial and management practices around the globe, but measures of their economic impact remain inconclusive.[7]

The Congress met in a context of looming crisis, as affairs in Germany and Italy demonstrated; Detlev Peukert famously described the years from 1924 to 1929 as a period of 'deceptive stability'. Despite currency stabilization Germany remained rent by cultural and political divisions, and, far from seeing rationalization as a salvation, many saw in it the cause of the nation's continuing struggle with unemployment. Unemployment undermined the bargaining position of labour unions; businesses objected to required funding for new unemployment benefits and had, by the late 1920s, become disillusioned with Germany's republican experiment. Italy was experiencing a decline in per capita GDP under a fascist dictatorship, and non-fascist labour unions had been crushed. Employers enjoyed new freedoms under fascism, and the government itself had sided firmly with scientific management, evidenced not only in Mussolini's remarks to the congress but in the establishment, in 1926, of a propaganda agency to promote the movement.[8] The context was one of increasing business and governmental authority alongside the weakening power of labour.

This was the context behind the six presentations analyzed below. Each comes from a prominent researcher, government official, or industrial or engineering manager, all invited to the congress because of their recognized specialist standing. Representing the American field of scientific management was Wallace Clark, prominent consultant and author, theorist of foremanship and management on the shop floor, and popularizer of Gantt charts, still widely used; and William Henry Leffingwell, stenographer turned office management theorist, also an author and consultant and dedicated proponent of the 'one-best-way', a phrase that has come to symbolize an authoritarian approach to effective management. Representing the rationalization movement of Europe were W. Pidgajetzky, member of the Ukrainian Academy of Sciences and collaborator with the Agricultural Research Institute of Kiev; Edmond Rognon, director of general operations for the transit system of Paris; Ludwig Ascher, medical officer of the Social Hygiene Research Institute of Frankfurt, consultant to Siemens electrical concerns and closely connected to Germany's *Reichskuratorium für Wirtschaftlichkeit* (Bureau of Efficiency); and Olof Kaernekull, consultant, chief engineer of the Industrial Bureau, founded by the Swedish Federation, and one of Sweden's leading advocates for rationalization. Kaernekull represented industry, while the other three Europeans held official posts or negotiated between official business and industry; the Americans were independent consultants. Kaernekull, Clark and Leffingwell attended the Rome meeting, although Ascher could not attend and mailed in his presentation. Whether Pidgajetzky and Rognon were in attendance in Rome remains unclear; Eleanor Cook reported that no official list of attendees was compiled.[9]

The analysis centres on the presentations as published in the *Acts of the Congress*.[10] The first and last days of the four-day conference were taken up with festivities, leaving only two days for presentations and discussion. The congress was divided into four sections: industrial, agricultural, service work and domestic work; some of the divisions had more than one hundred papers to get through. Attendees reported a hectic and raucous pace, with at most five minutes for presentation of an abstract, and five minutes more for shouted questions and discussion.

Performing Labour Management Expertise

What follows is an analysis of three ways management experts in Rome revealed the ideal that made their expert knowledge into expertise: the ideal of expert managerial authority over objectified labour. Contributors objectified labourers to varying degrees. Some identified workers by name, making their particularities integral to their own analyses and yet treating them as analogous to machines, so that their names became mere labels. Others made reference to workers but without identifying them in any particular way, through techniques that acknowledged the existence of personal particularities but did not reveal them. Still others obscured the personal particularity of workers entirely, by reducing their contributions to statistics. This analysis asks how experts in Rome treated the particular features that differentiate individual human persons one from another, as a way of tracking how real and specific working people came to be the objectified labourers who featured so prominently in the performance of managerial expertise.

Uniting differing expert treatments is a similarity in how the techniques on display performed. They simultaneously recognized and obscured the personality and character of individual workers. Gantt charts, photos of workers under observation, time-study sheets, and organizational statistics all performed a similar function: they argued that expert knowledge of work lay not with workers but with managers, and especially with managers who could deploy specialized administrative knowledge.

Common to all but one of the contributions analyzed below were methods of data collection. Expert knowledge in labour management required extracting information from competent workers by placing them under surveillance, and then gathering such information into general rules. Such knowledge relied on the competence of those under observation; competence meant that some workers would be good enough to serve as desirable examples and objects of study, and that they would be able to follow new management directives. Only the statistical work of Kaernekull departed from this method; Kaernekull observed business records rather than workers.

Common to all six was the insistence that management expertise would yield social benefits. The four researchers and consultants, Clark, Pidgajetzky, Ascher, and Leffingwell, wrote this explicitly. Kaernekull, a consultant but also a spokesman for Swedish industry, and Rognon, a French government official, assumed social benefits as a matter of course. Expected benefits included reductions in waste and better systems of distribution, which it was hoped would ameliorate problems of scarcity across Europe following the Great War. But none of the contributions dwelt on expected benefits, and the mood at the Congress was one of celebration rather than responding to dire social needs.

Identifying Workers

Gantt charts were a crucial feature of the American Wallace Clark's Rome presentation.[11] Clark was a specialist in the tracking of work and an avid user of Gantt charts; his enthusiastic championing helped bring them into widespread use. Mechanical engineer Henry Gantt, also American, had designed the charts to make a poor worker stand out like a sore thumb, and they remain a basic component of many management tool kits, from industrial relations to hospital management.[12] Foremanship was Clark's other specialty, and preparing and supervising skilled workers who took leadership on the shop floor.

Clark displayed a Gantt Man Record Chart in Rome, which identified particular workers by name.

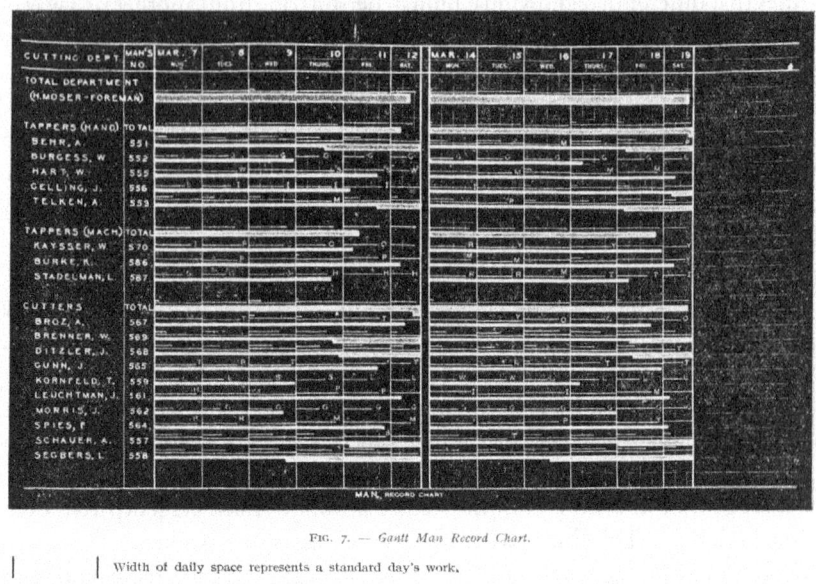

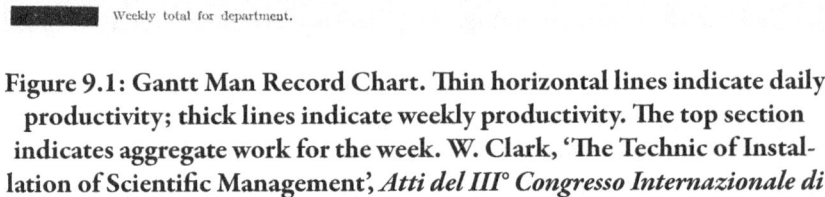

	Width of daily space represents a standard day's work.
	Amount of work actually done in a day.
	Weekly total of workman.
	Weekly total of group of workmen.
	Weekly total for department.

Figure 9.1: Gantt Man Record Chart. Thin horizontal lines indicate daily productivity; thick lines indicate weekly productivity. The top section indicates aggregate work for the week. W. Clark, 'The Technic of Installation of Scientific Management', *Atti del III° Congresso Internazionale di Organizzazione Scientifica del Lavoro* (Rome, 1927), pp. 449–57, on p. 455. Courtesy of the University of Minnesota Libraries.

Similar charts identified machines, and in fact the process started with a Machine Record Chart. Management reforms could not take hold unless shop foremen embraced them, and Clark argued that foremen were more likely to accept innovations in managing machines than in managing their workers. During an interview, a foreman was to be persuaded to admit that one of the most challenging tasks he faced was keeping his machines running when they were needed; the foreman was then enlisted in a joint project with upper management to quantify the time that machines sat idle. This is where the Machine Record Chart came in: it was a Gantt chart, displaying in dark, solid lines the time a machine was running, and leaving blank the times when it was not. Various reasons for machine idleness could be listed, and various steps taken to cut down on idle time, but the immediate point of the chart was to make visible the performance of each machine, so that corrective action could be taken.

The Gantt Man Record Chart used the same technique for tracking the performance of workers. Workers were organized by job and listed both by name (last name and first initial) and by worker number. To the right of their names were ten vertical columns indicating two five-day work weeks; thin horizontal lines represented work each day and thick lines represented work for the week. The top section represented the aggregate of work by all men for the week. Lines, thick or thin, meant good performance; short or missing lines meant failure.

Clark's Gantt charts thus identified specific working people, by name and work done, and by the types of delays they experienced; the charts reveal the people from whose competence Clark extracted his own expert knowledge. Workers' names were not incidental to the Gantt chart but integral in its interpretation; the chart compared workers with each other, and gave specific reasons for the slow work of any individual: G at the end of a short line meant the worker was new, a 'green worker'; H meant the work was too heavy, T meant trouble with tools, O meant troubles outside the shop.

These are identifying marks of individual persons, yet they are functionally analogous to assessments of machines. Introducing the charts into the workplace was a carefully planned manoeuvre, and required interviewing foremen, making analogies between machines and working people, and, finally, rewarding those who allowed themselves to be integrated into the new management plan. The Gantt charts revealed the raw human material from which observers extracted the data that allowed them, in turn, to assess the effectiveness of the method. The charts embodied, and indeed performed, the transition of workers from individual people to functions on a shop floor.

Referring to Workers

A number of contributions to the congress made reference to people, but without identifying them as individuals. Such references appeared in a variety of ways: in photographs of people as objects of study; in logs of the motions of named workers who nonetheless represented whole classes of labourers; and in the new availability of standardized but adjustable office equipment. Studies which merely referred to working people must be distinguished from studies which identified them; people are identified when their personal particularities become apparent because they are seen in the context of other people's particularities, not merely when they are named. A study may refer to working people by treating them as a class without providing the context of personal particularity that allows individuals within that class to be identified.

Several contributions to the congress in Rome belong in this category. The representative of the agricultural sciences division of the Ukrainian Academy of Sciences, W. Pidgajetzky, relied heavily on photographs of women at work in the fields in his report on the rationalization of sugar beet production.[13] His photographs came from the Research Institute for Agricultural Labour in Kiev, and they covered the whole process of the effort: from initial interviews with women tending beet fields to final photos of women, in regimented rows, performing rationalized labour. The study was a response to several concerns: a 1924 finding that sugar beet productivity had fallen below its 1913 level; eugenic concerns about the effects of hard field labour on women, especially during their child-bearing years; and general questions about the wear and tear of stooped labour.

Pidgajetzky's photos were full of women. The first (Figure 9.2) showed women being observed and interviewed by members of the Kiev research team, some sitting at rest, others bent over tools sunk into the soil.

Figure 9.2: Researchers observing women at work in the beet fields of Ukraine. Note their informal and non-regimented attitude. W. Pidga-jetzky, 'Problem der physiologischen Rationalisierung der Frauen-Arbeit auf den Zucckerreuben-Fieldern', *Atti del III° Congresso Internazionale di Organizzazione Scientifica del Lavoro* (Rome, 1927), pp. 415–20, on p. 416. Courtesy of the University of Minnesota Libraries.

Other photographs show the women after they had been trained in rationalized methods, standing still in a field in a straight row during a required break, and again, standing in a straight row demonstrating the proper length of the handle of a hoe. Pidgajetzky's colleagues in Kiev had found that requiring workers to devote 25 percent of their working time to rest yielded significant increases in productivity. The length of the handle of a worker's hoe also affected productivity, they argued; it should reach to her elbow and no further. This improved her performance and lessened her fatigue when she had to bend to the ground.

These are photographs of real women, marked by personal particularities, who nevertheless remain unidentified. Their stature varies, they wear different clothes, and where their expressions are visible they, too, differ from one another. Had such differences been judged of interest perhaps these women would have been identified, but neither Pidgajetzky nor his Kiev colleagues drew attention to these differentiating characteristics. It is possible for the viewer to identify some of the women by describing them, by referring to the woman with trim on her white skirt, or to the woman with her hair uncovered, but this is work the viewer must do herself. Such additional information was incidental to Pidgajetzky's report, and that it was incidental underscores the objectifying function of the expertise here performed. The variety that marks distinct human persons, which is apparent in the relaxed attitudes of Figure 9.2, were superseded by a regimented uniformity. It is that uniformity that embodied the expertise Pidgajetzky here performed; apparent individual differences such as in clothing or hair colour were incidental to his presentation.

Pidgajetzky was not alone in combining diagrams with photographs to preserve references to working people, while not identifying them. This technique illustrates one way in which expertise performed an objectifying function; specific and real people still appeared to the eye, but as representative figures rather than identified persons, and alongside their representations appeared more abstract figures representing the increasingly generalized knowledge extracted from observations of their labour. Three presentations illustrate this most clearly: Edmund Rognon's on scientific management at the Paris Transportation Agency, Ludwig Ascher's on measures to improve the efficiency and health of seated workers in German industrial and service sectors, and William Leffingwell's on scientific management techniques in the office.

A tension between diagrams and photographs illustrating human work was evident in Rognon's contribution.[14] Rognon, as general director of the Paris Transport Agency, was responsible for both its administrative and technical functions, and his presentation combined references to particular working people alongside schematic and abstract diagrams of the agency's widespread functions. The agency administered 124 tramways and eighty-five bus lines, a steam engine line of thirty-four kilometres and various boat lines on the Seine. Rognon presented an unusual round flowchart to depict general agency functions, and additional flow charts to illustrate relationships among various divisions. The final chart showed how tramway vehicles passed through the maintenance and repair system, and was followed by two photographs: the first of men standing alongside the tracks of the washing facility and awaiting a tram, the second of the same men now washing the tram; they had stepped in front of and alongside it and raised their long-handled brushes, but otherwise the scene remained the same. The effect is of a mimed sequence: the workers stepped forward and raised their brushes, almost in a salute. These photographs referred to people but did not identify them; they are photos of real people doing real work, but their function is general and objectivising: to represent the foundations upon which Rognon's expert knowledge is built, and to reveal how effectively his agency had eliminated interference from individual human characteristics.

Both productivity and health concerned Ludwig Ascher, medical officer of the Social Hygiene Research Institute of Frankfurt; he was unable to attend and mailed his presentation to Rome.[15] Ascher employed both photographs and diagrams, as had Rognon, but Ascher's diagrams were less abstract because they more closely resembled the photographs upon which they were based. Fatigue was the problem Ascher confronted, and he had helped to design workstations that minimized worker fatigue by providing support to the parts of the body that remained at rest. Skilled workers moved less, he had found; the hips and knees of an experienced blacksmith stayed pretty much in place, while a novice wobbled and swayed. The diagrams came first in Ascher's piece, as they had in Rognon's, although Ascher's were much less abstract; they depicted the paths of motion that workers had made, as caught on camera. A smooth, slim set of lines showed the skilled hammer strike of a joiner, his upswing etched right over the down-stroke; next to it was the wandering path of an apprentice, with a similarly smooth down-stroke but a wide and uneven swing on the way back up. Ascher's piece closed with four photographs of three different women in the electrical products industry, seated in supportive workstations following his designs. These photographs were Ascher's strongest reference to the particular people from whom he had extracted his principles of motion and support, but the women were not identified and appeared as representatives rather than particular persons. They embodied the ideal: workers fully encased within, and subordinate to, managerial expertise.[16]

William Leffingwell referred to people, without identifying them, in different ways than did Rognon or Ascher.[17] He presented no photographs, but used a variety of charts, tables and drawings. Office work was integral to scientific management for two reasons, he argued: because management work was organization work, and organization work was often largely clerical; and because clerical work was itself amenable to rationalization. Leffingwell had deep experience of office work having himself begun as a stenographer, and he was a founder of the American National Office Management Association and published widely on efficiency in office work.[18]

Two of Leffingwell's illustrations are especially revealing: the Time Study Sheet of Catherine Hailey in the department of General Sales of an unnamed firm, from observations taken on May 10 1927, by an A.C. Swanson (Figure 9.3); and a drawing of a chair recommended for office use, with an adjustable back rest and adjustable height.

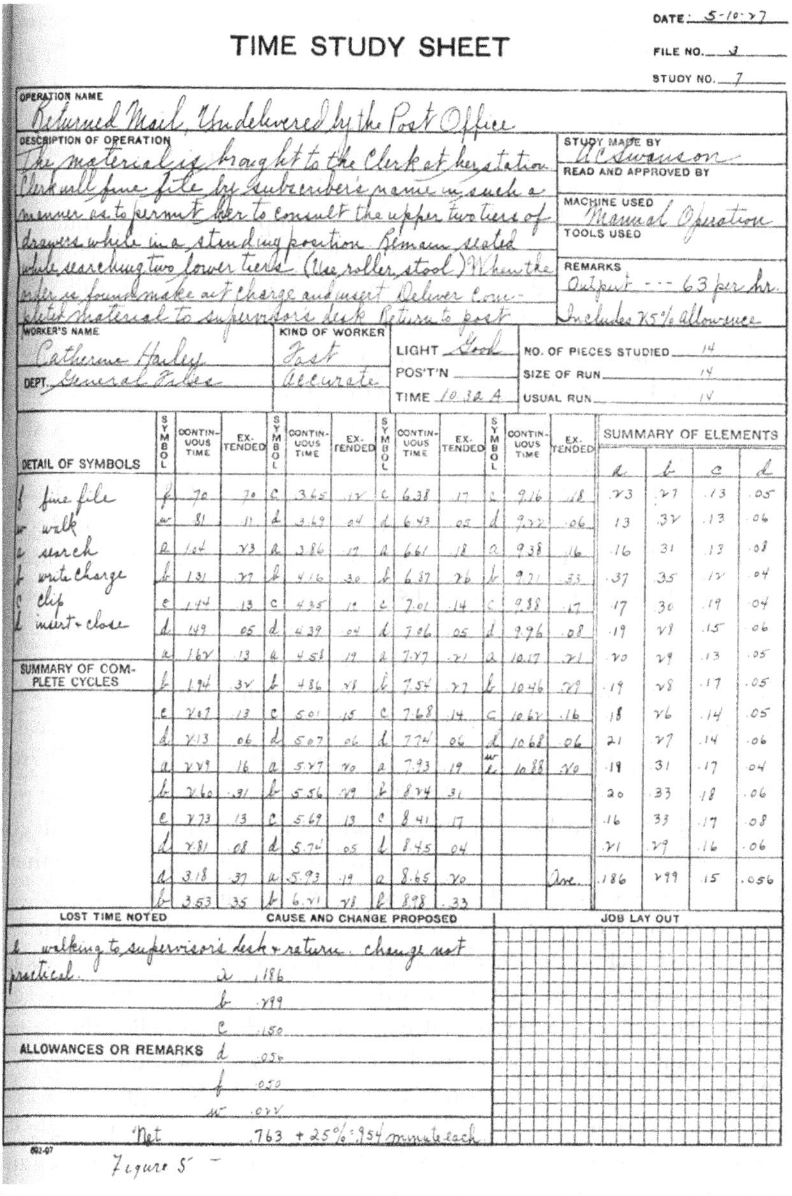

DATE: 5-10-27
TIME STUDY SHEET
FILE NO. 3
STUDY NO. 7

OPERATION NAME: Returned Mail, Undelivered by the Post Office

DESCRIPTION OF OPERATION: The material is brought to the Clerk at her station. Clerk will fine file by subscriber's name in such a manner as to permit her to consult the upper two tiers of drawers while in a standing position. Remain seated while searching two lower tiers. (Use roller stool) When the order is found make out charge and insert. Deliver complete material to supervisor's desk. Return to post.

STUDY MADE BY: A C Swanson
READ AND APPROVED BY:
MACHINE USED: Manual Operation
TOOLS USED:
REMARKS: Output --- 63 per hr. Includes 25% allowance

WORKER'S NAME: Catherine Harley
DEPT. General Files
KIND OF WORKER: Fast, Accurate
LIGHT: Good
POS'T'N:
TIME: 10.32 A
NO. OF PIECES STUDIED: 14
SIZE OF RUN: 14
USUAL RUN: 14

DETAIL OF SYMBOLS:
f fine file
w walk
a search
b write charge
c clip
d insert + close

SUMMARY OF COMPLETE CYCLES

SYMBOL	CONTINUOUS TIME	EXTENDED	SYMBOL	CONTINUOUS TIME	EXTENDED	SYMBOL	CONTINUOUS TIME	EXTENDED	SYMBOL	CONTINUOUS TIME	EXTENDED	a	b	c	d
f	.70	.70	c	3.65	.12	c	6.38	.17	c	9.16	.18	.23	.27	.13	.05
w	.81	.11	d	3.69	.04	d	6.43	.05	d	9.22	.06	.13	.32	.13	.06
a	1.04	.23	a	3.86	.17	a	6.61	.18	a	9.38	.16	.16	.31	.13	.08
b	1.31	.27	b	4.16	.30	b	6.87	.26	b	9.71	.33	.37	.35	.12	.04
c	1.44	.13	c	4.35	.19	c	7.01	.14	c	9.88	.17	.17	.30	.19	.04
d	1.49	.05	d	4.39	.04	d	7.06	.05	d	9.96	.08	.19	.28	.15	.06
a	1.62	.13	a	4.58	.19	a	7.27	.21	a	10.17	.21	.20	.29	.13	.05
b	1.94	.32	b	4.86	.28	b	7.54	.27	b	10.46	.29	.19	.28	.17	.05
c	2.07	.13	c	5.01	.15	c	7.68	.14	c	10.62	.16	.18	.26	.14	.05
d	2.13	.06	d	5.07	.06	d	7.74	.06	d	10.68	.06	.21	.27	.14	.06
a	2.29	.16	a	5.27	.20	a	7.93	.19	w l	10.88	.20	.19	.31	.17	.04
b	2.60	.31	b	5.56	.29	b	8.24	.31				.20	.33	.18	.06
c	2.73	.13	c	5.69	.13	c	8.41	.17				.16	.33	.17	.08
d	2.81	.08	d	5.74	.05	d	8.45	.04				.21	.29	.16	.06
a	3.18	.37	a	5.93	.19	a	8.65	.20			Ave.	.186	.299	.15	.056
b	3.53	.35	b	6.21	.28	b	8.98	.33							

LOST TIME NOTED — CAUSE AND CHANGE PROPOSED — JOB LAY OUT

l walking to supervisor's desk + return. change not practical.

ALLOWANCES OR REMARKS
a .186
b .299
c .150
d .056
f .050
w .022
Net .763 + 25% = .954 minute each.

692-07

Figure 5

Figure 9.3: Time Study Sheet. Note how closely observed were the subject's movements. W. Leffingwell, 'The Application of Scientific Management to the Office', *Atti del III° Congresso Internazionale di Organizzazione Scientifica del Lavoro* (Rome, 1927), pp. 5–20, on p. 15. Courtesy of the University of Minnesota Libraries.

Both illustrations appeared to refer to particular people, and even appeared to identify them: one was named, the other unknown but anticipated, in the allowance made for individual adjustments to the equipment. But neither identified the workers to whom it referred. The Time Study Sheet was attributed to Catherine Hailey, but it was not a record of her unique and particular work. Catherine Hailey was instead a representative of clerical workers generally; no comparison was made that might allow her particularities to appear, and any unique aspects of her performance do not figure in Leffingwell's analysis. Her name was simply a label. The Time Study Sheet gives ample evidence of the surveillance that was fundamental to rationalization and scientific management: Catherine Hailey was observed filing mail, from the firm to customers, that the Post Office had returned as undelivered. She did this sixty-three times in one hour, walking to the files, searching by subscriber's name, standing to use the upper two tiers of files and sitting on a rolling stool to consult the lower two tiers, writing out the charge, clipping it to the file, closing the file and delivering it to her supervisor's desk. She was described as a fast and accurate worker. The viewer might be tempted to consider Catherine Hailey as a unique and identified person, but she did not function as one in Leffingwell's work. What appear to be identifying details are akin to the uncovered hair or trimmed skirt of the unnamed women in Pidgajetzky's photos from the beet fields: possibly identifying information that a viewer can seek out, but included only incidentally, and not merely tangential to the analysis and aims of the Rome congress, but ignored. Embodied here again was the ideal of a worker stripped of individual characteristics, and open to the tools of expertise.

Because it could be adjusted, Leffingwell's chair suggests that it was itself representative of the individual and variable humans who would sit in it. The chair could be personalized, adapted to the specifics encapsulated in the archaic English term 'person': the unique and single body. But even here the particular person remained merely a reference: the body that would occupy this chair would function like any other body in the chair, no longer leaning forward and groaning with a backache, no longer swinging its feet because they didn't reach the floor. Leffingwell's chair is but one piece of a suite of rationalized equipment that embodied the desire to make irrelevant the unique variations that mark humans as individuals.[19]

Pidgajetzky's photos of women in Ukrainian beet fields, Rognon's photos of workers washing trams, Ascher's diagrams of bodies in motion, and Leffingwell's Time Study Sheet and adjustable chair all referred to people. They contained elements that call to mind recognizable human traits: a name, a deep seat for a human backside, real photos of real people. Despite these references to human characteristics, these contributions did not identify any people among those they studied or among those whose work they were intended to influence. In referring to people but not identifying them, these presentations contributed to the increasing generality and abstraction that marked the knowledge of the expert. Together they not only performed but celebrated the expertise of labour management, by showing it as effective and in force.

Obscuring Workers

The ideal that individual workers should be subordinate to managerial experts was most clearly embodied in Olaf Kaernekull's presentation. He made it immediately clear that his intent was not individual human performance, for record-setting results could easily be achieved in almost any type of task, he argued; instead he aimed to stimulate productivity through stable and long-lasting performance. 'Stars twinkle for a short time', he wrote – industrial operations required endurance.[20] Stars were single and individual, but industries relied not on single individuals but on steady, collective performance. Proof of Kaernekull's point came from an unidentified firm in Sweden's metal-working industry: it employed 1,100 people and made plates, tubes, wires and bolts out of copper, brass and aluminium. The company had reported its results with rationalization at a meeting of the Federation of Swedish Industry earlier in the year, having begun the effort in 1923.

The proof appeared in Kaernekull's statistics. Shipping, measured in kilogrammes per employee, had increased seventy per cent, total production had increased 24 per cent and the number of employees in the department had declined by 27 per cent, or by fourteen men. What happened to these fourteen was not clear; Kaernekull mentioned a rearrangement of work assignments without specifying what the new arrangements were or to whom they applied. Kaernekull also reported that rationalization methods had led to an 11.5 per cent decrease in the number of people working in the firm's rolling mills, while working capacity had increased by 13.5 per cent.

Kaernekull's statistics embodied an ideal of decreasing employment and augmented productivity; they had nothing to do with specific working people, although it is apparent that someone counted them. The benefits on display accrued at the level of the department and the firm; no mention was made of what they meant to those working in the firm, or to those who lost their employment. Not only did Kaernekull not mention any individual workers: he did not even refer to them. There is nothing in Kaernekull's contribution that calls to mind the recognizable human traits by which people identify or refer to each other as unique individuals. Here, people were obscured. This is what statistics do, and Kaernekull's contribution embodied an ideal of labour management expertise as an exercise in abstraction.

Conclusion

The congress in Rome was a celebration of the newly effective power of managerial expertise. Expert contributors performed demonstrations of the successful objectification of working people by displaying tools that functioned simultaneously as methods of surveillance and indices of success. Gantt charts and time-study sheets could only be filled up by managers who continually scrutinized their workers, and photographs and diagrams of proper posture and tool-use were by definition tools of watching. Charts, study sheets and photographs also yielded evidence of how successful managerial experts were, primarily by documenting worker compliance; few provided evidence of wider social and economic benefits.

The objectification of working people can also be seen in the context surrounding the congress: fascist, anti-labour Italy as its host, in an atmosphere of looming international crisis. But a close reading of expertise as performed in Rome provides a more fine-grained analysis of the steps by which objectification was, and may be, achieved. It provides a way of considering the effectiveness of expertise, by revealing ways that methods of study can also function as their own justification. We would do well to continue broadening our understanding of expertise, by examining how it is constructed and functions outside liberal societies, and also to focus our understanding more sharply, by looking at the specific contents of expertise in its many forms.

We would also do well to consider when expertise may cease to be susceptible to negotiation and mediation, or when it may become congealed or hardened. The expertise on display in Rome was an expertise of discipline, designed to ensure that workers produced a desired result. It was designed precisely to prevent worker interference, even unintentional interference through fatigue or poor training. In other words, it was an expertise that had congealed and was no longer open to negotiation with workers. Workers were objectified and obscured, through the performance of an expertise based on surveillance and founded not on trust, but on discipline.

And so we join again the delegates to the congress at the Campidoglio, among the soldiers on the stairway to the Senate, and consider again the trumpet blast that announced the arrival of Mussolini. Working people were struggling; their organizations were eclipsed by business and government, and scientific management, with its uncompromising dictate, was not on their side. The trumpet blast may have signalled victory, or it may have sounded a charge; either way its reference was to war.

10 SCIENTIFIC EXPERTISE IN CHILD PROTECTION POLICIES AND JUVENILE JUSTICE PRACTICES IN TWENTIETH-CENTURY BELGIUM

Margo De Koster and David Niget

This chapter focuses on the influence of experts and various forms of scientific expertise on the Belgian juvenile justice system and child welfare policies in the twentieth century. Within this specific context – the institutions devoted to the disciplinary re-education of 'criminal' and 'deviant' subjects – expertise takes on the particular form of a procedure which introduces the authority of science into the exercise of state power, through complex technological modalities of risk assessment, but also through the philosophical primacy of reason and its relation to truth as a form of legitimization. The 1912 Belgian Child Protection Act is interesting in this respect, since it introduced the instrument of social- and medical-psychological inquiry into the judicial treatment of delinquent youth. Not only did this reflect a significant shift in the problem-definition of juvenile delinquency, it also created an important new, formal and institutional 'space for experts' within the juvenile justice system.

This development was part of a larger movement in which several forms of expert knowledge served to formulate a new set of doctrines and ideas about child welfare policies. In the first part of this chapter, we discuss how firstly judicial and criminological knowledge, then medical, psychological and pedagogical knowledge and, finally, sociological knowledge entered reform debates in this field.

In the second part of the chapter, we try to assess the actual role allotted to, and taken up by, expert diagnostic practices and power within the Belgian child protection-cum-juvenile justice system. We examine the organization and importance of 'scientific' observation of juvenile delinquents or 'child guidance' within the central state observation centres for boys (Mol) and for girls (Saint-Servais) from 1913 to the 1950s.

We aim to demonstrate how the practice of expert diagnostic observation of 'problem children' contributed first to a medicalization and later to a psychologization of both the definitions of juvenile delinquency and the responses of disciplinary re-education. Further, we argue that through this performance of expertise[1] and the prism of this specific expert gaze, expertise came to play a crucial role in the construction of juvenile justice cases and the shaping of juveniles' individual trajectories through the system. At the same time, however, we try to indicate that the experts' power and their disciplinary discourse did not reign supreme, but rather involved 'negotiation' with other forms and actors of power: it remained largely confined to the inside of the juvenile justice system and was effective only insofar as it met the ambitions and interests of juvenile justice officials on the ground and could be integrated into broader child protection discourses.

Finally we shall show that even if, later in the century, certain expert knowledge and practices – such as psychology and sociology – tended to promote giving a degree of autonomy to young people in conflict with the law through a discourse of individualization and responsibilization, it has to be observed that the power of the expert kept these young persons in a subaltern position, more objects than subjects of expertise.

The Expert as Policy Reformer Judicial and Criminological Expert Knowledge

The turn of the twentieth century was marked by the emergence of a new penal doctrine, influenced by the rise of criminology. This was the doctrine of 'social defence', elaborated by the Belgian penal expert, Adolphe Prins, at the end of the nineteenth century in the wake of the German legal expert, Franz von Liszt, with a dual concern for socialization on the one hand and for 'subjectification' of classical penal theory on the other.[2] It reflected a paradigm shift from a liberal concept of law that evaluates criminal facts and sanctions them in proportion to their gravity, to a preventive concept linked to the identification of the criminal and his or her treatment. In this context, where reducing the risk of (further) criminal behaviour was prioritized over the punishment of crimes and misdemeanours, children and adolescents became the legitimate and privileged targets of a 'predictive' and socially efficient penal intervention, aimed at hindering the development of criminal behaviour.[3]

This doctrine of 'social defence' paved the way for the input of expert knowledge from the new science of criminology, based on the study of the 'criminal individual'. An influential person in Belgian criminology, Louis Vervaeck, a licensed medical practitioner, established *Laboratories of Criminal Anthropology* in Belgium's prisons from 1907 onwards.[4] Influenced by the Italian and French schools of criminal anthropology, he implemented a standardized and 'scientific' procedure of medical, psychiatric and social evaluation of prison inmates, which would serve as a model for the observation institutions for young delinquents. His objective was both curative and repressive: amendable delinquents had to be cared for and trained to be useful to society, the 'incorrigible' ones had to be 'eliminated' through confinement of 'undetermined' duration.[5] His son, Louis Vervaeck, continued his father's work by becoming 'inspector of the asylums and medical-psychological institutions', which the observation centres for delinquent young persons were part of. He was able to feed the Belgian child protection system with new theories and institutional innovations from abroad, both Anglo-American and French.[6] This clearly shows that Belgium stood at a crossroads of Germanic, Anglo-Saxon and Latin influences, making it a haven for institutional innovation.

Medical, Psychological and Paedagogical Expert Knowledge

The 1912 Belgian Child Protection Act, which instituted juvenile courts, was part of this new penal rationality. This law replaced the notion of 'discernment' (the ability to tell right from wrong) with an idea of the educability of the individual offender. Thus, the new law mandated a preliminary study of the environment and personality of young delinquents, with input from a 'social inquiry' on one hand, and a medical and psychological examination on the other. This examination was to be carried out in 'child guidance' or observation institutions: here, the new child behaviour experts would soon find a privileged status within the legal system, and be established at the institutional level.

The technique of 'observation' was influenced by the new science of 'paedology' as a comprehensive 'science of the child' that came into being in Europe from the late nineteenth century onwards as an attempt to create a study of children's behaviour and development in the manner of natural sciences, based on a positivist methodology.[7] The approaches of anthropometrics, psycho-physiology and paedo-physiology were privileged. Soon, these were complemented with approaches and insights from the psychological, paedagogical and educational sciences in the blossoming field of the scientific study of children and adolescents.

Parallel with the work of Alfred Binet in France,[8] Edouard Claparède in Switzerland[9] and William Healy in the United States,[10] numerous public and private initiatives were taken in Belgium in order to build up a coherent body of knowledge of child psychology. The leading figure of this network was Ovide Decroly, a licensed neurologist and international expert in the field of educational science in the early twentieth century. After setting up special schools for 'abnormal' children in Brussels, which were serving as models across Europe, he extended his expertise to juvenile delinquents.[11] At the first child protection conferences, he stunned legal experts and criminologists by stating that their 'understanding of the criminal act and crime was still very simplistic'. He called for the development of real scientific expertise and the rational and empirical classification of 'irregular' children, as well as individualised treatment.[12] (Brussels: Moniteur belge, 1913), pp. 159 and 182. In 1920, he opened a day clinic in Brussels, which

Brussels's juvenile judge Paul Wets quickly came to consult on a regular basis.[13] As Depaepe, Simon and Van Gorp have demonstrated, Decroly was both a laboratory man, pioneering the application of psycho-technical tests and clinical methods, and a networker, capable of penetrating savant circles (conferences, associations, committees etc.), building relationships and fuelling debate.[14] He was also able to involve himself with public policies, while at the same time being critical towards them. He was, thus, the ideal type of what an expert should be.

In the field, it was Maurice Rouvroy, previously a schoolteacher, who came to play the role of architect of the institutional observation of juvenile delinquents. After becoming the director of the Central State Observation Centre for boys in Mol in 1913, he acquired national and then international recognition, even though not a university graduate, which also proves the impact of practitioners in this movement.

With the opening of the Mol observation centre in 1913, the practice of child guidance or 'observation' of young delinquents appeared early in Belgium, as soon as in the US, in Chicago.[15] Mol quickly served as a model for other European countries. A similar centre for delinquent girls was set up in Saint-Servais, a suburb of Namur, and began operation during the First World War. In the second part of this paper, we will see that child guidance, which was presented as a diagnostic tool for individuals, quickly turned out to be of great importance in the management of the juvenile justice system.

Sociological Expert Knowledge

Finally, besides this initial criminological and psychological-educational movement, the interwar period saw the emergence of sociological expert knowledge, under the impetus of a leading figure in twentieth century child welfare in Belgium, Aimée Racine. A practising lawyer, member of the Solvay Institute of Sociology,[16] and former post-doctoral fellow in criminology in the USA, Racine was well-informed about American debates on juvenile delinquency. Working closely with Paul Wets, the juvenile judge of Brussels, she conducted an extensive field study in the 1930s, temporarily taking on the role of juvenile probation officer, and published the results in a detailed monograph on juvenile justice children and their social environment.[17] After the Second World War, she became the head of the new Centre for Research on Juvenile Delinquency, attached to the Free University of Brussels.

In her work, Racine downplayed the importance of medical-psychological expertise, insisting more on the merits of social and educational fieldwork that involved the families of 'problem' children and youth.

Not surprisingly, this development of sociological knowledge went hand in hand with a move towards professionalization of child protection actors, with the opening, in the 1920s, of the first schools of social work.[18] However, this professionalization only began to have a significant effect on actual practices towards the end of the 1940s, when the recruitment of professional juvenile justice staff (probation officers) was imposed by law.

The early twentieth century thus saw the development of various forms of expert knowledge around the figure of the 'problem child'. We observe here the introduction, within the field of justice – a sovereign power – of the process of 'scientization' of the state and of public policies which had been under way since the nineteenth century. Science, especially in the form of expert knowledge, conferred a new authority on the power of the state, that of governing based on the holding of knowledge.[19] Although distinct, sometimes even conflicting rationalities were involved, the child protection reform movement provided a new space in which different experts could meet and share interests. In this way, child advocacy can be seen as a movement able to federate different professional agendas and interests.

Expert Diagnostic Practices and Power within Belgian Juvenile Justice

The two Belgian state observation institutions successfully occupied the new formal space for experts created by the 1912 Child Protection Act. They quickly came to constitute an important hub in the juvenile justice system, a systematic and influential partner of the juvenile judges and the administration of the juvenile justice penitentiary institutions.

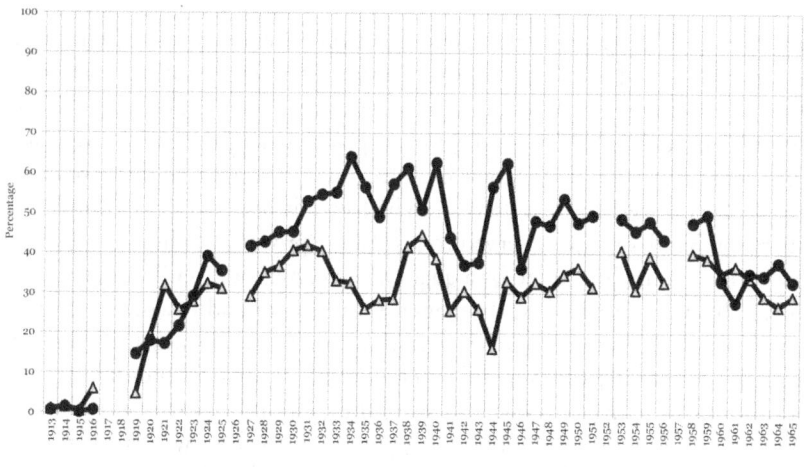

Figure 10.1: Ratio of the number of placements in observation to the total number of minors tried. Public and private institutions. (Source: Judicial Statistics of Belgium, 1912–65)

From the 1920s onwards, placement in observation institutions became increasingly popular among juvenile judges and came to represent 30–40% of boys' sentences, and 40–60% of girls' sentences, involving between 800 and 1,500 individuals per year.

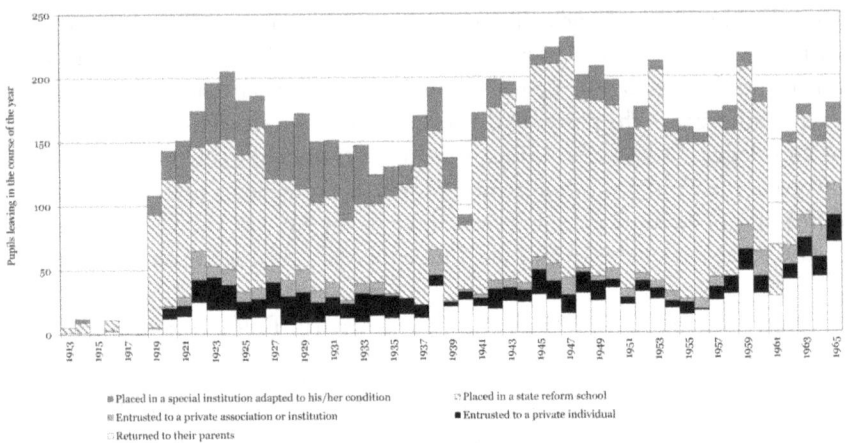

Figure 10.2: Reasons for detained persons leaving the Saint-Servais State Observation Institute by decision of the juvenile judge or the minister of justice. (Source: Judicial Statistics of Belgium, 1912–65)

Another interesting point is the tendency of juvenile judges to sentence these juveniles who passed through observation more often to institutional placement (rather than other possible re-educational measures, such as provisional liberation under probation) than others, and to placement in the most severe state reform schools, renowned for their harsh regimes. It seems like the outcome of the expertise realized during observation had already been anticipated and set in stone beforehand, or merely served to justify the privation of liberty – hypotheses which are both confirmed by a qualitative analysis of juveniles' case files.

Overall, the impact of expert diagnostic practices and discourses on juvenile justice practice should not be overestimated. There was no emergence of a so-called autonomous 'medical and psychological jurisdiction' monopolizing all decision-making within juvenile justice; on the contrary, expert power was more fragmented and contested than at first appearance. To begin with, both the official *Judicial Statistics* and individual records of juveniles placed in observation suggest that, beyond their role of expertise, the observation centres also served as a safety valve in case of organizational problems and overcrowding of other juvenile justice reform institutions.[20] In addition, although juvenile judges were in most cases inclined to follow the conclusions of the observation report and the experts' advice on the appropriate re-educational measures,[21] the influence of these recommendations is rather equivocal. The psycho-medical expertise was generally welcomed by the judges precisely because it provided them with 'scientific' arguments demonstrating the need to pursue a process which they had already set in motion. Following on an often very detailed social investigation which had already allowed the judge to form an opinion, the psycho-medical and paedagogical expertise came to confirm the direction given by the latter – in this case, the placement in a state reform institution.[22]

Overall, within the juvenile justice system, the experts in the observation of 'problem children' not only presented themselves as successful 'technicians', but were also perceived as such, because their expertise generally answered this system's need to measure, classify and successfully organize the flow of juvenile cases through the juvenile penitentiary institutions, and because it gave a more solid ground to the decision-making process of the juvenile judges. This implies that these experts exerted a great deal of power within juvenile justice practice, but only insofar as they stuck to the expert role that juvenile justice officials allotted them and did not overstep their prerogatives.[23] Indeed, in general, although juvenile judges (in particular the Brussels juvenile judge, Paul Wets) greatly contributed to establishing the reputation of the state observation centres as 'jewels' of the Belgian child protection system, they did not want the penitentiary administration and its experts to interfere with their work. When this did happen, the administration's decisions were openly contested by the judges and even caused major conflicts. In 1919, for example, Huy juvenile judge, Derriks, declared of the Mol observation centre:

> This school is closed to us, its board organizes all by itself scientific experiments with the classification of children whom we put at the disposal of government. And then it makes of the children what it wants. It pretends ... to replace us.[24]

Neither the role and power of the experts, nor the actual conclusions of the expertise developed through either the medical-psychological or the paedagogical observation of juvenile justice children went, therefore, uncontested. Moreover, throughout the period from the start of this expert diagnostic practice in 1913 up until the 1950s, the expertise in itself did not constitute a homogenous or static disciplinary discourse, but rather reflected a shifting power balance between various disciplines and techniques playing a part in it: from the medical specialists to the psychologists; from religious personnel to professional paedagogues and educators; from the distant observer to persons in intimate contact with the psychoanalysts. In the following sections, we take a closer look at these evolutions and their impact on the nature and outcome of expert diagnostic practices.

Mol and Saint-Servais between the Wars: the Medicalization of Delinquency

The mission of the Mol and Saint-Servais observation centres was threefold: first, to carry out observation as part of a psycho-medical examination of individual juvenile delinquents; second, as a result of this observation, to prescribe the most appropriate treatment or re-education regime for each individual case; and third, to develop a classification of juvenile delinquents based on scientific criteria.[25] The observation centres soon became real juvenile delinquency study centres; the steadily growing number of boys and girls being scrutinized produced a vast pool of scientific data, making it possible to present annual reports on the causes of juvenile delinquency.[26]

There were slight differences between the treatment of boys in Mol and that of girls in Saint-Servais, but overall one can state that the systematic and daily observation of all pupils constituted the basis of both institutions' scientific activity. The Mol institute was regarded as a real 'paedological' research laboratory; in Rouvroy's words 'a psycho-paedagogical clinic', because observation and test results directly informed the re-education of the pupils.[27] In Saint-Servais, on the other hand, medical and moral evaluation predominated, with elaborate psychological testing only applied much later, owing to the lack of professional training of the nuns who carried out the observation of the girls.

The observation of pupils in Mol and Saint-Servais was organized in two phases: medical-paedological examination and paedagogical observation – a very innovative approach at that time. Each newly arrived child was transferred to a 'life-group' in the pavilion system, and within this small group of pupils living together he/she was observed and evaluated as to his/her potential for being re-educated. During this diagnostic observation a complete personal and medical record (medical, somatical, neurological, psychiatric) of the youngster was drawn up, using an extended medical checklist to establish a complete case history. The child's reading and writing abilities were tested, and as a first entry into the paedagogical study of the child's personality and mental life, inmates were asked to make a drawing or write a few lines on an autobiographic subject such as 'the greatest joy' or 'why I am unhappy' or 'my life'.[28] During the second phase in the family system, based on the idea of self-government and the model of the school-city system, constant observation by all the members of the educational staff was the rule, in order to – to quote Rouvroy – 'penetrate the soul of the child'.[29]

The medical and paedagogical observation was completed by a whole range of diagnostic tests.[30] Besides the classical Binet-Simon intelligence scale used to measure children's mental age, use was also made of tests developed by Decroly, Buyse and Vermeylen, together with techniques such as aesthesiometry (the measurement of the degree of tactile or other sensibility) and algesimetry (the measurement of sensitiveness to pain).[31]

This practice of intensive testing leaves some doubt as to the actual use of the numerous test results in diagnosis. Aimée Racine, for example, criticized the incoherent multiplication of tests which, in her eyes, did not reveal anything about the more profound causes of juveniles' misbehaviour.[32] One can assume that, on the one hand, the laboratories equipped with sophisticated machinery served in the first place as showcases to impress the numerous visitors of the observation centres, foreign experts and practitioners in particular. On the other hand, the testing ritual which juveniles underwent during their observation contributed to their symbolic subjection to the authority of science, without truly nourishing the diagnosis.

Another remark to be made is that, besides brand-new scientific knowledge, the observation yielded results that came very close to nineteenth-century perceptions of youth crime and its causes. Although an ever-growing number of medical, psychiatric and psychological concepts, categories and explanations appeared in the observation reports, in the end, the overwhelming majority of juveniles were classified as 'morally disordered' children. Juvenile delinquency thus continued to be defined, in the first place, as a moral problem, the causes of which were invariably sought in poverty, family instability and bad influences of the social environment and the street. Here, scientific expertise revealed its common-sense roots.[33]

The Postwar Period and the Psychologization of Expertise

The post-war period was one of institutional change in the area of child protection in Belgium. From 1947 onwards, the Saint-Servais staff was made up of professionals: trained social workers, both religious and non-religious, who acted as 'educators', a professional 'psychologist' and a 'psychologist's assistant', and a doctor, specialized in psychiatry. The new head of the Saint-Servais observation centre even held a PhD in educational sciences, obtained from the University of Leuven in 1934.[34]

At that time psychology enjoyed great legitimacy, and the discipline would gain great influence on juvenile justice practices and procedures. Reflecting the rising influence of this paradigm, the so-called 'progressive' system whereby girls were placed in separate sections according to a merit system based on their morality was replaced in 1949 by a division into age groups, following intelligence quotients and professional competences.[35]

The development of psychology produced a paradox, however: while promoting a comprehensive approach and strengthening the idea of responsabilization of inmates ('self-government' and importance of subjectivity), the use of projective techniques, such as Rorschach tests, tended to de-responsibilize the individuals, whose troublesome behaviour was seen to be rooted in deep emotional imbalances beyond the individual's control.

Although psychology enjoyed great legitimacy, one can observe the persistence of a medical approach towards the inmates, which even increased with the appearance of anti-psychotic drugs in the late 1950s.[36] In 1959, a 'special section' was opened at Saint-Servais for the most 'difficult' girls. In an almost systematic way, high doses of tranquilizers and antipsychotic drugs were administered to them. The swift adoption of this particular scientific discovery in juvenile reform institutions marked the return to a disciplinary order and a very deterministic interpretation of behavioural disorders in terms of pathological corporeality.[37]

Overall, the experts advocated responsibilization and enhanced autonomy of the inmates, but in practice this was jeopardized by the influential and deterministic approaches of heredity and psychiatry, as well as by the later development of psychology: these were obviously techniques of knowledge and power which kept boys and girls in a subordinate position within the institutions, as if they were objectified by expertise. The persistent weight of psychiatry, aided by the invention of neuroleptics, and the influence of the institutional disciplinary system, explain how young people were kept in this subalternate position, despite injunctions, in the post-war period to promote the autonomy of the poorly adapted young persons.[38]

The Role of Expert Reports in the Construction of Cases

The production of these medical and paedagogical reports transmitted to the juvenile courts consisted of a reordering of observation notes, test scores, conclusions from the social investigation, medical results etc., which also served to give meaning to the often lengthy judicial 'career' of the young delinquent under observation. The construction of this narrative testifies as much to the perspective of the observer as it does to that of the observed.[39] Through the selection of information, the interpretation of signs as symptoms and finally the causal ordering of biographical elements from the young persons' lives based on techniques of narrative writing, the medico-paedagogical expertise report not only reveals its categories of thought, but also constructs a 'case'. It seems that the narrative provides the ability to link together the scattered clues provided by various psycho-technical tests. Above all, it builds a bridge between the visible and the invisible, between the traces of disorder suggested by the environment and the behaviour of the juvenile delinquent, and their 'real' influence on the young person's personality. In the end, this narrative provides the link between diagnosis and prognosis, whereby the reconstructed past is projected onto a likely future scenario, which is explicitly staged in recommendations for treatment and re-education which constitute the epilogue of the observation report. On the epistemological level, the case analysis does reflect an attempt at individual diagnosis – this being its official function – but it is also processed, in turn, as a global resource for practitioners, who then draw from these cases their empirical knowledge. Moreover, many of these become 'case-studies' put forward in annual reports, comments by directors of the institutions, sometimes also in scholarly publications. As a result, there is a circularity of knowledge between these individual case files and the practice of medical-psychological and paeda-gogical observation, whereby this process of collecting information from the biographies of the young observed is constitutive of the accumulative knowledge of experts, a knowledge which is then applied as a power to young people. [40] It is in the interaction between individual cases and universal knowledge that resides the performative dimension of this detailing of biographic facts by experts. The 'clinical' case is registered, compared with other cases by analogy, inserted into a typology of differentiated cases, and finally referenced against an exhaustive body of knowledge which pretends to universal validity.

Conclusion

It is in this apparent contrast between the empowerment of the youngsters and the constraints and power imposed on them by the expertise, that we need to seek its particular nature and role in the field of government policies of social regula-tion and punishment – more precisely its political rationality. The involvement of experts in state regulatory policies made it possible, as Nikolas Rose has indicated, to articulate liberalism and authority in the exercise of power within modern liberal democracies.[41] Indeed, liberalism created a number of problems for the gov-erning of societies, by setting limits to the use of power, while at the same time the spontaneous regulation of social problems proved to be difficult, if not impossible. Hence the crucial role played by expertise in state intervention in the social sphere. Expertise made it possible to shape individual liberty in the collective framework of society, and to justify a 'conduct of conducts', i.e. the conjunction of solidarity and socialization policies with the promotion of the modern, autonomous and responsible citizen. Science, especially in the form of expert knowledge, conferred a new authority on the power of the state – that of governing based on the holding of knowledge. 'Scientific' practices of measuring and classifying social problems were increasingly believed to produce more adequate and effective responses to social problems and risks. New expert knowledge on crime and disorder, drawing heavily on the medical sciences in particular, was thus allotted a key role in the cor-rect 'diagnosis' of problems, thereby legitimizing the 'civilizing' and disciplining responses of the state to citizens' disorderly behaviours.

In this context, childhood has offered particularly fertile ground for expert knowledge, constituting an epistemological resource for the quickly developing sciences of psychiatry, psychology and paedagogy. In the early twentieth century, a whole army of experts devoted itself to the fate of the child and the adolescent, thereby creating, and drawing from the child's vulnerability, troubles or danger-ousness, a new area of social intervention in which they could operate. Whether in the fields of public or mental hygiene, or even penal prophylaxis, new forms of knowledge – medical, psychiatric and psychological, or sociological and criminological – all oriented public policy towards childhood, considered both as an epistemological matrix and a powerful resource for social intervention by the state, because childhood was considered as a tool to articulate the temporality of risk. The field of child protection and juvenile delinquency thus became a privileged site for the 'crystallization' of the larger process of the 'scientization of the social' from the nineteenth century onwards.

The case of expertise on juvenile delinquency is particular in this sense that it presents a power relationship between the experts and their silenced 'objects'. This relationship did not only result from the embeddedness of experts in an institutional disciplinary system, sanctioned by law, but was also shaped by the nature of the expert knowledge, rooted in the academic disciplines of medical science, psychiatry and psychology. These disciplines all informed a predominantly pathological interpretation of behavioural 'disorders' in terms of mental, physical and psychological deficiencies, and thus provided techniques of knowledge and power which kept boys and girls in a subordinate position, as if they were objectivized.

Although in individual cases, experts in the field of juvenile justice were not always successful in implementing their particular views in reality, and sometimes met with outright hostility, their expertise was certainly effective: in practice, it actively served – and was mobilized by juvenile justice actors – as a tool for the legitimization of judicial intervention and for organizing the case flow within the justice and penitentiary system (the observation centres as 'hubs'). Thus, the experts of juvenile justice had considerable impact on the judicial decision-making process concerning the trajectories and treatment of the youths involved. Underlying this overall success of expertise was not the exact content and quality of the expert knowledge produced, but rather the (almost ritual) performances of 'scientific' testing, measuring, observing (the testing equipment as expertise's 'front stage') and producing observation reports ('back stage'). These performances, in particular the testing ritual which juveniles underwent during their observation, contributed to their symbolic subjection to the authority of science.

11 EXPERTISE AND TRUST IN DUTCH INDIVIDUAL HEALTH CARE

Frank Huisman

A silent revolution seems to be taking place in western health care systems. In the Netherlands, this became clear in 2009, when medical lay people challenged the authority of the profession and the state twice, using the Internet as their weapon. In two instances, the general public was warned against the dangers of vaccination in general, and of the vaccines on offer in particular. One was against HPV – the virus that may cause cervical cancer – the other was against H1N1, the virus responsible for the 'Mexican flu'. Both vaccination campaigns were supported by the highest medical and political authorities. In both cases, however, scientific claims of experts were cast in doubt and their authority was challenged.

What is the meaning of this? How is it possible that the medical profession and the state have come to be doubted? Has there ever been an era in which the profession and the state were trusted? If so, when and why? These are important questions, which seem to be at the heart of this volume. But let us first take a closer look at developments with regard to the HPV vaccination campaign.[1]

The human papillomavirus (or HPV) is known to cause abnormal cell growth of skin and mucous membranes. In the case of an infection, there is an increased chance of developing specific forms of cancer, among them cervical cancer. It is estimated that in the Netherlands, 200 women die from cervical cancer every year. In 2008, two vaccines became available: Gardasil (developed by Merck) and Cervarix (produced by GlaxoSmithKline). They were said to decrease the risk of developing cervical cancer by 70 per cent. The Health Council advised Ab Klink, then minister of health care, to include the vaccine in the National Vaccination Programme, meaning that vaccination would be offered for free by the state. Expectations were that the vaccine was capable of protecting women against the HPV virus – which is sexually transmitted – provided that they had not been sexually active yet. The minister agreed, and he decided to provide the vaccine for free to girls of 12 years and older.

A huge informational campaign was launched by the National Institute of Public Health and Environmental Hygiene (RIVM), which was supported by the Ministry of Health, the Health Council, the Dutch Society of General Practitioners (NHV), the Municipal Health Services (GGD) and the Dutch Vaccine Institute.[2] A website and a newsletter were developed, and folders, brochures, posters and key rings produced. Finally, there was a special phone number that people could call to report complications. In terms of information and organization, nothing was left to chance. However, when vaccination started in March 2009, it became clear that the information campaign had not really caught on. The turnout was much lower than was to be expected based on earlier experience. In advance, it was thought that more than 80 per cent of the girls summoned would show up. In the end, only 45 per cent completed the whole series of three vaccinations.[3] What had happened?

Doubt with regard to vaccination had been created on the Internet. Stories were circulating claiming that the vaccine was genetically manipulated and even life-threatening. The low turnout was mainly attributed to the Internet campaign that the Dutch Society Critical Vaccination (NVKP) had launched. The society had been established in 1994 by people with bad vaccination experiences.[4] It argued that citizens were not informed about the potential dangers of vaccination, while research into harmful side effects was delayed because of the financial interests of pharmaceutical companies and the career interests of biomedical researchers. On its website, the society complained about the great social and medical pressures with regard to vaccination. They were considered to threaten citizens' autonomy and freedom of choice. It was claimed that education and screening could work just as well as vaccination, and should therefore have priority.

One could consider the website of the Dutch Society Critical Vaccination as an attempt to educate the general public by informing it. But on the Internet, there is much more on offer. For lay people it is difficult to distinguish between 'good' and 'bad' information – which in liberal democracies is a matter of attribution by definition. Another website, hosted by an organization called Niburu, claimed to supply 'awareness-building news'.[5] With regard to the HPV vaccination, it 'revealed' a plot of the authorities to counter overpopulation. It was suggested that the vaccine contained nanochips that would cause a slow but certain death. The campaign was said to be coordinated by the Global Alliance for Vaccines and Immunization, which was a 'global health partnership' in which the Rockefeller Foundation, the Bill Gates Foundation, the United Nations, the World Bank, UNICEF, the World Health Organization and many western governments were said to participate. Whoever doubted that this was actually happening was referred to witnesses such as a top official in the American army, a Cambridge professor and a professor affiliated to the French Centre National de la Recherche Scientifique.

Again: what is the meaning of this? Is this just an isolated example of paranoid confused minds? Or does it point to a broader trend in modern society?

It has been argued that today, science in the service of the common interest is threatened, as scientists and policymakers have come to see science mainly as a servant of interest group politics.[6] This is by no means an academic issue, interesting only to STS scholars (who work in the field of science, technology and society studies) or to historians of science. It is a concern to medical practitioners as well. To quote from an editorial published in the authoritative *British Medical Journal*:

> Today, clinical reality as perceived by clinicians has to be reconciled with patients' beliefs, 'resources' have to be balanced against individual patient need, and ethical dilemmas spring hydra-headed from medical advance ... Utterly unquestioned biological givens are disintegrating all around us ... Doctors will become purveyors of choice – or agents of control – within the plastic limits of the flesh ... To the postmodernist the question is whose 'evidence' is this anyway and whose interests does it promote?[7]

Have we lost trust in the sources of expertise? Or did we gain awareness about the way expertise and trust are constructed? Whereas some celebrate the triumph of the autonomous citizen-patient who has finally liberated himself, others deplore the erosion of professionalism, arguing that our national health care system is under threat. Big issues are at stake. They include truth, professionalism, political leadership, responsibility, distributive justice, trust and citizenship. Are we entering a new phase in the history of expertise?[8]

The case of HPV is interesting because it incorporates two levels of health care and responsibility (collective and individual), two patterns of morbidity (infectious and chronic) and two paradigms of coping: vaccination and lifestyle. From the late nineteenth century onwards, the state had taken ever more responsibility for collective health by putting sanitary measures in place.[9] This had led to a dramatic fall in mortality caused by infectious disease, especially after antibiotics and vaccination had been introduced. However, the second half of the twentieth century witnessed a shift in morbidity pattern from infectious to chronic disease.[10] After the 1940s, infectious diseases had ceased to be the dominant killers. Due to lifestyle changes (smoking, eating and drinking behaviour), chronic diseases like cancer and cardiovascular disease were on the rise. Because these were related to lifestyle, they could not be prevented or cured by 'magic bullets' supplied by the state, but only through changes in lifestyle, suggested by expert advice on causality patterns. Whereas accepting expertise became increasingly problematic in the democratic knowledge society, this was especially true in the case of HPV vaccination, in which a 'quick fix' (using the logic of infectious disease) was used to counter the risks of sexual behaviour (belonging to the domain of chronic and lifestyle disease).

This chapter is looking at the relationship between the medical expert, the citizen-patient and the state in The Netherlands, taking the complexities into account which were outlined above. I will be looking at three moments in the history of Dutch medical expertise in individual health care, a domain which is characterized by a different dynamic than public health. While in public health, there is a need for technocratic expertise, in individual health care patient autonomy and trust are essential drivers. In the field of individual health care, the choice of the citizen-patient for a specific healer is related to the attribution of expertise by the patient. It is dependent on factors like credibility, trust and persuasion, rather than on power, discipline or coercion, as in public health. As the editors argue in their introduction, the 'expert society' was never fully within the hands of the presumed experts. Expertise was always a socially constructed, inherently unstable form of authority, only materialising through the *performance* of experts. I will argue that this especially applies to the field of individual health care. Here, it is not so much a matter of technocratic implementation of scientifically informed blueprints for society, but rather of trust and negotiating scenarios. In the following, I will present three stages of the development of Dutch individual health care: legislation (1865), resistance (1913) and reform (1993). Over the course of time, this did not necessarily lead to 'progress' through the doings of medicine and the state. Today, modern western health care is characterized by all kinds of contradictions and paradoxes.[11] In the final section of this chapter, I will suggest three ways of coping with these paradoxes.

Experts, the State and Society

In order to put developments in individual health care into perspective, we will have to look at developments in public health first, because the two have always been related in complex ways. As the literature on the topic makes abundantly clear, the relationship between experts and the state has never been unproblematic.[12] First of all, it would be a mistake to think that expertise consists of a homogeneous body of knowledge waiting to be accepted and implemented by the state. Secondly, once experts and the state have agreed on a scientifically informed policy, it has never been a matter of simply imposing a blueprint on a passive society. Citizens could either ignore, resist, change or accept policy measures, but responses were never predictable or straightforward.

This also applies to public health. In a classic and influential article on state responses to epidemic disease in the nineteenth century, Erwin Ackerknecht suggested that there is a relationship between political orientation and prophylactic regime.[13] During the nineteenth century, Europe and its colonies were struck by many serious epidemic diseases, including yellow fever, the plague, typhus and the biggest killer of them all: cholera. While thousands of people suffered and died from these diseases, their causes remained unclear. Physicians and states felt an urgent need to understand their aetiologies and come into action.

Generally speaking, there were two theories available to explain the causes of epidemic disease. The first was contagionism, according to which a *contagium vivum* is transmitted from person to person. In order to prevent contagion, it made sense to isolate the sick. Because quarantines presupposed state power and the will to use it, this theory was appealing to autocratic regimes. According to the second theory – anticontagionism or miasmatic theory – it was the environment which was pathogenic. The environment produced so-called *miasmata*, which were propagated through the air rather than by persons. Quarantine – considered to be a despotic measure – was not only seen as damaging commerce, but even as useless in countering epidemics. The only preventive against epidemic disease was progress of civilization. Sources of *miasma* included overcrowding, filth, dampness, faulty drainage, vicinity of graveyards and unwholesome water and food. Against all this, quarantines were useless. Instead, sanitary measures were called for. The debate between both camps was never just a medical one, but always a debate on state intervention as well. Hence, while the leading contagionists were high-ranking military or navy officers, leading anticontagionists were known radicals or liberals, keen on sanitary reform.

The Ackerknecht thesis on the connection between political ideology and preventive policies has been highly influential, informing the work of many medical historians.[14] The attraction of his presentation of things is in the clear and dichotomous nature of health policies: while autocratic regimes feel attracted to contagionist policies, liberal ones tend to adopt miasmatic theory, leading to sanitary policies. However, attractive as it may be, this image is too simplistic. The focus of Ackerknecht was on the interaction between experts and the state. He was thus ignoring society, implicitly suggesting that citizens passively accepted prophylactic measures once experts and the state were agreed.

The image Peter Baldwin is giving in his comparative study of the response to three contagious diseases (cholera, smallpox and syphilis) in four European countries (Germany, France, Great Britain and Sweden) is much more complex, precisely because he is including society.[15] He starts by establishing that the main challenge of any state is to find the proper balance between the interests of the community and those of individual citizens. Protecting society against the potential threats of infectious disease implies an infringement on the autonomy of individuals by definition. But quarantine measures, compulsory vaccination or regulation of prostitution need to be socially sanctioned and politically legitimized. Secondly, Baldwin argues that a strictly binary view of aetiology and prophylaxis is a distortion of reality. Local factors (whether natural or social), individual predisposition and contagion all played their role – they were mutually permeable. As a result, prophylactic strategies employed by European states have always been very different. Quarantine was not simply a matter of political authorities forcing measures on a passive population. More often than not, the authorities acted because they felt the pressure of public opinion in favour of a quick solution. Very often, contagionism was the common-sense aetiology of the average person, whereas environmentalism – with its attribution of disease to unseen factors and its bourgeois insistence on salubrity and personal hygiene – was learned behaviour.[16] In his book Baldwin seeks to explain why prophylactic responses to similar epidemiological challenges were so different across Europe. The answer is that responses were always dependent on specific contextual factors like political philosophy, geo-epidemiological location, perceived risk, commercial interests and administrative capacity. Last but not least, preventive strategies against contagious disease went to the heart of the social contract.

This introduces the important dimension of public trust, which is especially relevant in the field of individual health care. Trust has to be earned and it may be lost. It is by no means self-evident that citizens consult an academically trained physician when they fall ill. In this chapter, I will be looking at three moments in the history of Dutch (individual) health care. First, I will be looking at the consequences of national legislation on the medical profession introduced in 1865. Formally speaking, academic professionals had earned a monopoly of treatment. In reality however, they needed to work very hard to earn credibility and trust in a society that was hardly aware of the new situation. Next, I will analyze the debate following a petition to liberalize these laws in the 1910s. Public awareness of the implications of the laws of 1865 had grown, and an important group of citizens felt the need to contest it. To no avail, because the intervention state was on the rise. Finally, I will discuss the implications of current legislation, promulgated in the 1990s, and implying a liberalization of legislation. The welfare and intervention state was on the decline and citizens were increasingly defined as well-informed patient-consumers moving on a transparent market.

Medical Legislation of 1865: The Liberal State

In 1865, the Dutch Parliament accepted four laws concerning the organization of the medical profession and the national health care system.[17] This legislation was to frame and regulate Dutch health care for well over a century. The laws can be considered as part of the nation-building process, in which the patchwork of local professional competences was replaced by one integrated national system. In the early modern period there had been academic *doctores medicinae*, barber surgeons, apothecaries, oculists, herniotomists, *maître-dentistes*, cutters for the stone (operators removing bladder stones), medicine vendors, midwives and many others. Their training had been at either one of the European universities, or in a local trade guild or through the practice of life; their legal competence had been either for a specific town or for the countryside, the army or the navy – or they were consulted because they just happened to be there.

Johan Rudolph Thorbecke (1798–1872), the minister of internal affairs who had drafted the laws, was keen on creating one unified medical profession and one unified pharmaceutical profession; on raising medical, surgical, obstetrical and pharmaceutical teaching to academic levels; on making unauthorized medical practice (i.e., without academic title) liable to punishment and on introducing a National Health Inspectorate. Thorbecke expected the laws of 1865 – together with the Law on Higher Education that was to follow in 1876 – would raise the 'level of study and competence' of medical practitioners.[18] Henceforth, the only possibility to become a physician or a pharmacist was to enroll in the medical or pharmaceutical programme at one of the four Dutch universities. For that reason, his legislation was welcomed by the Dutch Society for the Advancement of Medicine (NMG) and its counterpart, the Dutch Society for the Advancement of Pharmacy (NMP), as well as by leading people in the medical and pharmaceutical fields.

Academically trained medical professionals had fought hard for these laws to materialize, and expectations were that they had given the profession a social mandate – or even a monopoly. This, however, was hardly the case. At best, the Thorbecke laws had created a formal framework within which the medical profession could mature and grow. After 1865, there was much work to be done in terms of organizing trust and credibility. In this respect, little was to be expected from the state; it was all up to the profession itself. On the ground, physicians and pharmacists were working hard to make a living and survive. It was up to the leaders of the profession – the professors at the university and the executives of the professional bodies – to create public trust. Let us take a closer look at the implication of the laws of 1865 for the social standing of pharmaceutical experts. Many believed that the laws had granted pharmacists a monopoly in the field of the preparation and delivery of medicines.[19] It was expected that the druggist-trade would gradually disappear, creating full professional opportunities for academic pharmacists. This, however, did not happen. Indeed, it had never even been the intention of Thorbecke. According to this liberal politician, the state should limit itself to supplying medical and pharmaceutical education of academic standards, and to the inspection of the professional conduct of the academic professions. Consumer choice or interprofessional relationships were, however, no concern of the state. The implication of this was that pharmacists had to earn credit and legitimacy with the public on their own. They had to define (and uphold) their academic identity and professional standards while at the same time fight a tough struggle for survival with many competitors in the medical marketplace, including dispensing physicians, druggists, quacks and the emerging pharmaceutical industry.

In drawing up his bills, Minister Thorbecke had been led by three guiding principles: first, he wanted to separate pharmacy (*artsenijbereidkunde*) from medicine; second, he wanted to enhance the position of pharmacists by allowing those who were qualified to study pharmacy at university and finally, he intended to limit the competence of druggists to the wholesale trade in medicines. Henceforth, whoever wanted to practise pharmacy had to meet stringent requirements. On top of that, there were strict laws regulating pharmaceutical practice. A Medical State Inspectorate had been installed to supervise compliance with the law. In 1876, the Law on Higher Education stipulated that pharmacy and toxicology were to become academic disciplines. Each of the four Dutch universities got its own chair for pharmacy and toxicology. Henceforth, the only road to pharmaceutical competence was through university.[20]

Expectations were that this would raise pharmacy to academic levels. There was, however, a long way to go before this situation would be realized: first, most pharmacists had not (yet) received an academic education and second, patients were hardly inclined to value the pharmacist more than other medicine vendors. Even after 1865, trust had to be organized. It was crucial to create a public image of pharmacists that fitted in with the high ideals of their professional leaders. The newly appointed professors of pharmacy devoted themselves to this task. In their inaugural address and in deontological publications they created a professional image of the pharmacist as a man of honour and science, who was entitled to a prominent position in the health care system.

Willem Stoeder, one of the four newly appointed professors of pharmacy in Amsterdam, published a series of articles meant to contribute to a new representation of pharmacy and its practitioners. In this series, called 'Letters from the Capital', Stoeder developed a deontology for the modern pharmacist.[21] He pointed out that the pharmacist was no mere shopkeeper but a scientist, who should try to earn public esteem and patient trust. Being a man of honour, he had been called to solemn duties. Therefore, he should not tarnish his reputation by selling secret remedies or non-pharmaceutical commodities like paint, perfumery and the like. In this context, Stoeder contrasted the 'fairground attraction' of the shop-windows of Paris pharmacists with the 'tasteful simplicity and dignity' of their German colleagues. Social success would be the logical result of scientific education and ethical elevation, Stoeder argued. The other newly appointed professors also tried to remove current prejudice about pharmacists by creating a new professional image. The traditional pharmacist had been a mere shopkeeper, who had been subordinate to physicians because he acted on their instructions, written down in prescriptions. He was seen as a medicine mixer, as someone putting together active substances. The modern pharmacist, on the other hand, was a man of science; a man with independent views and the equal of physicians.

To what extent did pharmacists live up to the professional profile their scientific leaders had created? Evidence suggests the 'field' had less elevated thoughts on the topic. In order to become legally competent as a pharmacist, the state exam of pharmacy sufficed. Only very few pharmacists used the opportunity to take their doctoral degree of pharmacy.[22] Out of 130 persons who had become pharmacists since 1876, only three had valued the doctorate of pharmacy enough to take their degree.[23] Pharmacists did not seem to consider the doctorate a prerequisite for the practice of pharmacy or, for that matter, as a means to increase their social prestige. Although the medical laws of 1865 had consider-ably raised requirements for entering the profession, the material conditions and social prestige of pharmacists had not improved proportionally. Among other things, this was caused by the enormous expansion of the pharmaceutical indus-try, leading to the replacement of many galenic remedies by chemical ones.[24] Patient demand for manually prepared galenic remedies was declining, whereas manufactured synthetic medicines became a booming business. Secret remedies and *spécialités* were introduced to the market in increasing numbers and varie-ties – which is to say nothing of the competition by physicians and druggists. Mutual infringements by physicians, pharmacists and druggists on each others' professional domain abounded, and quackery was thriving.

Long after the medical laws of Thorbecke had come into effect in 1865, phar-macists were being warned by the Medical State Inspection, or even fined in court, for dispensing medicines without prescription, for the sale of secret remedies, for the unauthorized practice of medicine, for not complying with the provisions regarding stocks, pharmaceutical weights and scales, for having unauthorized locums in case of absence or for giving the key to the poison cabinet to unauthor-ized persons. Although it is very difficult to establish the scale on which these violations took place, it can be established that the professional consciousness of the pharmacists in the field had not developed as their leaders would have wished. In 1873, the Medical State Inspectorate established the fact that secret and other remedies were being sold by unauthorized persons 'almost everywhere'.[25] In the annual reports, mention is being made of the sale of a whole range of medicines by pharmacists' assistants, veterinarians, merchants, housewives, a clergyman, a photographic dealer, a saddler, an innkeeper – in short: by everyone.

In the liberal political climate of the late nineteenth century, pharmacists had to steer a middle course between the demands of science and those of the public. Whereas science was held in high esteem at modern university, the public had different standards. This made the legislation of the years 1865 and 1876 highly Janus-faced: while pharmacy had become a scientific discipline, the profession had hardly won in terms of credibility and social acceptance. The support of the patient was not won by the image of the pharmacist as a man of science and high virtue, simply because it was unaware of these public relations campaigns. Pharmaceutical experts and the general public were living in different worlds. Many patients preferred the remedies of unauthorized medicine vendors to those of academic pharmacists. Maybe the best way to look at the laws of 1865 is to regard them as an ideal – drafted and supported by the medical and political elite – of which Dutch citizens were hardly aware. During the decades following legislation, this awareness was gradually growing. In 1913, a petition was sub-mitted to the Dutch Parliament, arguing for a liberalization of the legislation of 1865. It was supported by almost 8,000 citizens.

A Petition in 1913: The Rise of the Intervention State

In 1913, three Dutch lawyers submitted a petition to the Dutch Parliament, in which they requested a liberalization of the 1865 legislation.[26] They included Samuel van Houten (former minister of the interior), Joost Adriaan van Hamel (professor of criminal law at the University of Amsterdam) and Rudolph Otto van Holthe tot Echten (councillor of the court of justice in The Hague).[27] Because they had serious doubts about the competence of physicians, they disputed their exclusive right to medical intervention. Medicine and health care should serve the interests of the patient rather than those of the physician, they argued. The petition caused much commotion in Dutch society: as many as 7,700 people expressed their approval of its contents, while it prompted Dutch government to ask the Central Health Council and two State Committees for formal advice on the issue.[28] During the years in which the petition was under discussion many articles, brochures and pamphlets were written, both in favour of it and against it.

The three lawyers decided not to be distracted by complex epistemological matters, but to focus on the legal dimensions of the issue instead. Thus, they established the fact that the law of 1865 was violated on a daily basis, which they considered to be an infraction of their sense of justice. At the time, it had been the intention to put an end to the confusing patchwork of training facilities and legal competences in medicine by allowing only academically trained physicians to practice medicine. However, this goal had not been accomplished – quite the contrary. The High Council of Justice (*Hoge Raad*) had had to step in quite often to issue jurisprudence in interpretation of the law.[29] A situation of 'total legal insecurity' had come into being. The three lawyers proposed to end this situation by adapting legislation to current practice. Secondly, the three claimed that the law of 1865 denied citizens the right to decide on their own fate. They wondered what vital state interests were threatened when unlicensed healers came to the rescue of patients, and they called for legal guarantees ensuring the freedom of choice of citizens in case of illness.[30]

As long as it had not been proven that health matters were better served by physicians than by others, that freedom of treatment in individual health care would cause damage to public health and that legislation could be an effective weapon in the battle against unlicensed practice, it was impossible to maintain the monopoly of treatment for physicians. The three lawyers called for an amendment of the 1865 legislation to allow patients to seek treatment from the healer of their choice. The state should limit itself to regulating the training and examination of aspirants to the profession. Further, the state should take legal action against conscious deceit of the public and against any speculation on its ignorance. Finally, the petitioners wanted *all* malpractice to be prosecuted, regardless of whether it had been committed by official physicians or by unlicensed practitioners. What they called for, in short, was a fair political balance between narrow professional interests and the general interest of citizens.

During the parliamentary debates in January 1914 feelings were mixed.[31] Some MPs felt that implementing the petition would move the health care system in a dangerous direction, pointing to the evils done by quacks. Others argued that as a matter of principle, the state did not have the right to restrict the freedom of choice of citizens in a matter that was as personal as individual health. For the time being, the Dutch government withheld its point of view. The minister in charge was Prime Minister Pieter Wilhelm Adriaan Cort van der Linden.[32] Being a lawyer and a liberal, he thought that the issue was both fundamental and complex. Although he granted that the state should not act as guardian vis-à-vis its citizens, he thought it unacceptable when a liberalization of current legislation would endanger public health. He decided to consult the Central Health Council, the government's supreme advisory board in health matters.

In its report, the Health Council confirmed that illegal medical practice was indeed thriving. It could not be denied that the law's major objective had not been accomplished. Indeed, according to the Health Council the law of 1865 was 'nearly a dead letter'.[33] The council was facing a thorny problem: on the one hand, helping fellow human beings in distress should not be liable to prosecution and punishment. There was, however, ample evidence of irregular healers causing damage to people's health. Therefore, the Health Council advised to organize an impartial investigation of healing systems unexplained by science. Should it turn out that inexplicable forces or faculties existed, the government should seriously consider taking steps in order to alleviate human suffering. In that case, many legislative changes were called for.

In line with this advice, two State Committees were established. They were asked to look into the legal and the medical dimensions of the matter. The Legal State Committee was made up of six lawyers, with Van Houten serving as chair.[34] Very soon, the committee presented its advice.[35] The Van Houten Committee proposed to introduce the principle of 'limited exclusive competence'. It wanted everybody offering medical help to register with the health inspector of their district. Healers were expected to specify the sort of treatment they were offering. Secondly, so-called 'restricted interventions' were defined, in which only regular practitioners should be permitted to engage – notably surgery, obstetrics, the prescription of strong medication and the treatment of venereal disease. Finally, everybody would henceforth be held legally liable for the effects of a medical treatment. With its proposal, the committee hoped to dispel three objections against existing legislation: the proposed bill would be more in line with people's overall sense of justice because it fitted better with existing realities in the health care system; it would honour the principle of patient autonomy and self-determination and finally, it would encourage irregular healers to use their gifts in the interests of mankind.

Cornelis Pekelharing, professor of pathology in Utrecht, chaired the Medical State Committee, which consisted of medical professionals.[36] It had been assigned to study the value and effects of unorthodox healing methods. The committee invited 'all who believe they qualify' to come to Utrecht, allowing the committee to assess their therapies. Several healers contacted the committee, bringing along one or more patients. In total, ninety-six patients were prepared to cooperate. The committee concluded that the healers made poor or even wrong diagnoses; that the phenomena and methods they worked with were well known to physicists and physiologists (and could therefore not be attributed to unknown, mysterious forces), and that their treatment results were quite limited and in some cases even negative. Having studied the potential merits of magnetism, Christian Science, somnambulism, herbal medicine, homeopathy and naturopathy, the committee concluded that there was no evidence of healing methods leading to new or surprising cures. Its final conclusion was that the results 'in no way substantiate the view of those who feel that it is in the interest of mankind to recommend the practice of medicine without scientific training'.[37] The law of 1865 should remain in force – a conclusion that was warmly welcomed by the medical establishment.[38]

Initially, the medical community had not taken the petition of the three lawyers very seriously. However, when it became clear that the prime minister was prepared to give it some serious thought, the need for a clear collective response was felt. The editorial board of the *Dutch Journal of Medicine* decided to devote a special issue to the matter.[39] This issue, entitled 'Medical monopoly and freedom of healing' (*Artsenmonopolie en geneesvrijheid*), contained twenty-six contributions written by twenty-three authors. Space forbids going into the arguments put forward by them in any detail, but overall they claimed that science was superior to intuition, and that healing practices should always be rooted in science. 'Science' was used as a demarcation criterion between official and unlicensed healers. Although this seemed to be a clear criterion, it merely led to additional questions, such as: how should (good) science be defined? And: how does scientific knowledge inform medical practice? In his introduction to the special issue, Gerard van Rijnberk – professor of physiology and editor-in-chief – made an attempt. Medicine, he argued, included 'all knowledge, tested by experience and experiment, of the material structure, the workings and the defects of the human body, as it is gathered through the ages and taught at university'.[40] Several essays looked into the basic characteristics of science. While some emphasized the open character of medicine, others felt that its defining characteristics were to be found in system and knowledge. Either way, medical science was considered superior to lay knowledge. Van Holthe tot Echten, one of the three lawyers, summarized the reasoning of physicians as follows: 'having knowledge is better than having no knowledge at all and we are the ones who own medical knowledge'.[41]

Because the petition touched on the boundaries of law and medicine, it was impossible to settle the dispute with arguments derived from either of these disciplines. In the debate on freedom of healing the protagonists could only fall back on general cultural notions of body and mind and of health and illness. It was the very general dimension of the issue that constituted the petition's broad appeal from the outset. That is why 7,700 individuals felt the need to support it by signing it and that is why there was such extensive coverage in the general press.[42] Still, it is important to establish that the petition movement was no medical counter-movement, and did not intend to be. Although the content of the petition specifically reflected a critique of medicine, it was aimed at an overall mentality that merely engaged with the material, while disregarding the immaterial, the spiritual, the intuitive and the mystical.[43] The petition movement, it seems, was primarily motivated by epistemological doubt. These wider implications of the petition were not lost on the Central Health Council. In its advice to the minister, it indicated that the subject matter was 'highly complex'.[44] It was up to politics to weigh the many contradictory feelings, considerations and arguments involved.

The petition had a political dimension in that it raised the issue of the relationship between state and society, and of the role that experts could or should play in individual health care. There was perhaps no one who could defend the petition's political dimension with more authority than Van Houten.[45] As a down-to-earth rationalist he had never engaged in the humanitarian-idealist experiments of his time. Still, he was very committed to the 'liberty' and the 'natural rights' of the patient.[46] He considered Thorbecke's legislation as an 'unauthorized interference by state authority' and felt that punishing unqualified healers was indefensible because the initiative in consulting them lay with patients. Moreover, Van Houten – who had great confidence in science – saw the physicians' passionate response to the petition as a sign of insecurity concerning their own therapeutic powers. When these were indeed so limited, there could be no good reason to protect them using the force of law.

Although Van Houten and Prime Minister Cort van der Linden had rather different views on liberalism and the role of the state, they tended to agree on the issue of freedom of healing. A liberalization of the 1865 legislation seemed within reach. However, things worked out differently. Much time had passed since the submission of the petition in 1913. When the State Committees finally presented their reports, it was too late for Cort van der Linden to act on them. In the elections of 1918 – the first after the introduction of general male suffrage – the liberals suffered a very considerable loss. Their number of seats in parliament dropped from forty to fifteen. A new confessional Cabinet, led by the Catholic Charles Ruys de Beerenbrouck, came into office.[47] The new Cabinet felt no need to spend much time on the issue of liberalizing medical legislation. Precisely in the area of social legislation and state intervention the Ruys de Beerenbrouck

Cabinet was very ambitious.[48] The 1865 medical laws simply remained in place and the issue disappeared from the political agenda. It may be speculated that the young nation state felt no need to deconstruct institutions that had just been put in place. Changing things for individual health care may have put public health structures in danger as well, and the new Cabinet must have been wary of that implication. The net result was that the law of 1865 was still violated frequently, and that alternative and unlicensed healers remained active – much to the annoyance of the society against quackery. This points us to a discrepancy between legal competence and practical trustworthiness – or, to put it differently: between the attribution of expertise by the legislative state and patients in day-to-day reality.

Individualising Care in 1993: The Neo-Liberal State

Eighty years after the petition was submitted, its major tenets proved politically convincing after all. In 1993, the Law on Professions in Individual Health Care (Law on PIH) was enacted. In many respects this new law reflected the proposals that had been advanced by the Van Houten Committee. The law's two underlying principles were freedom and protection: patients should henceforth be completely free in the choice of their care provider and they should be protected against professionals' negligence or deceit. This double objective was to be achieved by abolishing the prohibition on the practice of medicine by alternative healers, by setting up a register for the protection of training and title, by formulating so-called 'restricted interventions' and by amending criminal law and medical disciplinary rules.[49] The new legislation was almost a one-on-one implementation of the proposals put forward by the three lawyers in 1913. What had happened in the meantime?

Over the course of the twentieth century, the cultural authority of medicine had risen sharply in the Netherlands – as in the wider western world – due to dramatic improvements in the field of diagnostics and therapeutics. Penicillin, kidney dialysis, radiotherapy, polio vaccination, open heart surgery and even heart transplants and the CAT scan are among the breakthroughs of medicine.[50] Combined with the general improvement of living conditions, average life expectancy at birth had almost doubled in less than a century. During the postwar years, science was held in great social esteem. There was a general trust that the world could be made safe and prosperous by building on science. In the 1960s and 1970s, however, people began to seriously doubt science, when it became clear that there was another side to it, like pollution, mass destruction and iatrogenic mistakes. The notion of progress through science was cast in doubt, while the use of scientific experts was criticized by intellectuals like Thomas McKeown, Michel Foucault, Ivan Illich and David Armstrong. The 1980s and 1990s witnessed the crisis of the welfare state.[51] During these decades, it became fully clear that governmental care from the cradle to the grave was no longer financially feasible. In the Netherlands, drastic cutbacks were realized. The state withdrew itself from many domains, leaving them to the dynamics of the market. In the domain of health care this implied that medical professionals were replaced by health managers, while discretionary competence was replaced by third party accountability.[52] Similar trends took place in the rest of the western world, with the British Prime Minister Margaret Thatcher and the American President Ronald Reagan as the embodiment of the global neo-liberal revolution, which also affected health care systems.

In this political climate, the legislative state proved to be sensitive to the criticism, taking steps to protect patients against the nearly unlimited power of physicians. The law of 1993 can be seen as the culmination of a broad liberalizing movement that began to unfold in the late 1960s and led to a legislative boom aimed at regulating the legal position of patients vis-à-vis medical experts. The doctor-patient relationship was regulated in a so-called medical treatment agreement between the patient and the care provider. Henceforth, the client was invited to select the care provider of his choice, with whom he would enter into a business-like relationship. The client could discontinue a treatment agreement at any time, without reasons given. Care providers were henceforth required to inform patients on diagnosis, prognosis and therapy, and they were not allowed to act without the patient's consent. Finally, care providers could be held liable for their treatment, according to civil law, criminal law and disciplinary rules. Henceforth, the patient was considered to be a well-informed citizen, moving in a transparent health market. The citizen-patient had grown into a consumer-patient.[53]

Many physicians deplored this 'juridicisation' of the health care system. They felt that medical intervention cannot be reduced to a legal contract. Indeed the patient's position has been strengthened, but it is questionable whether this juridicisation offers a solution to the paradox of modern life. The irony of living in a democratic knowledge-society is that we are no longer capable of shaping an opinion on our own, while we are constantly challenged to do so. How, for instance, should we evaluate the decision of Dutch TV-personality Sylvia Millecam, who sought treatment for her breast cancer with a faith healer called Jomanda? This case – which has grown into a *cause célèbre* in The Netherlands – seems to embody all the paradoxes of modern health care. When the lump she felt in her breast turned out to be an operable tumour, Millecam decided not to visit an oncological surgeon. The faith healer she consulted instead denied that she was suffering from breast cancer. It remained untreated, the tumour grew and two years later Millecam died. The Dutch Parliament and the Health Care Inspectorate used her death as an occasion to reopen the debate on the 1993 law.[54] Jomanda and three regular physicians, who had all treated Sylvia Millecam, were brought to trial. Two of the physicians were suspended for life from medical practice by the medical disciplinary tribunal, while Jomanda was acquitted. The law of 1993 had created two domains in the field of individual health care: one in which academic doctors were active, for whom the rules of university training, academic practice and strict inspection applied, and one which may be called 'the free medical marketplace', where citizens could freely choose healers whose competence was not guaranteed by the state and who were not subject to the criteria of the Health Inspectorate.

This leads to the following questions. In a liberal democracy, should patients be protected against unorthodox healers by state-enforced law? Or should the principles of bodily autonomy and self-determination apply? If modern secular individuals are expected to make well-informed decisions, how can they be sure that they fully oversee the consequences of their actions? And then there is the issue of liability in case something goes wrong. Although the 1993 legislation seems to provide a clear frame, it can be quite difficult to apply in everyday care practice. For instance, the 1993 Law on Professions contains no explicit norms as to the meaning of notions like 'professional standard' or 'care of a good care provider'. In case of a dispute between a doctor and a patient, the judge acts as the arbiter who is expected to formulate an expert legal opinion on what constitutes expert medical practice. It is one of the ironies of history that the 'liberation of the patient' goes hand in hand with a call for even more expertise.

Epilogue: Organizing Trust

It has been argued that the more people know about the workings of science and technology, the more realistic their expectations with regard to the problem-solving capabilities of experts become.[55] In other words: the more information about science is available, the more people realize that experts are fallible. In this sense, the 'crisis of 2009' can be understood as part of a process of increasing transparency and increasing reciprocity between state and society; a process that has been taking place since the late-nineteenth century. In many ways, the HPV debate reflects a broader challenge to modern health care structures.[56] In the late twentieth century, western societies witnessed a transformation from the intervention state to the neo-liberal state. In the process, citizens came to be defined differently.[57] Whereas over the course of the twentieth century health care had become a matter for the state and access to it became a civil right, towards the end of the century, citizens came to be held responsible for their own welfare and health. They were invited to behave as informed consumers, moving in a (supposedly) transparent knowledge society. In the process, citizens learned to be critical – or even distrust – both experts and the state.

The so-called new public health is related to the rise of risk calculation and management as ways of dealing with growing uncertainties in technologically advanced and deregulated liberal societies. Considering everybody at any moment as a potential patient, predictive and preventive medicine focuses on risk profiles in relation to factors such as age, class, occupation, gender, lifestyle and consumption. Interventions take diverse forms, such as periodic check-ups and genetic testing of groups at risk for specific illnesses, public education about 'risky' lifestyles and skills development.[58] The 'new public health' is all about providing individuals [*sic*] with information about their health status and possible health risks, so that they can act to reduce those risks. At the same time the risk discourse does not provide certainty. Scientific and expert knowledge on health risks is intrinsically provisional. Also, it gives cause for disagreement: not only among experts, but between expert and popular views as well. Medical information is increasingly located on the free market, where competition and various players with different interests are involved: medical researchers, public-health experts, clinicians, epidemiologists, the pharmaceutical industry and patient organizations. So again: did we lose trust in experts? Or did we just gain awareness about the way expertise is constructed? Either way, how should trust in experts be organized in our neo-liberal, postmodern era? This is a daunting question, if only because of its inherent paradox: articulating the issue of trust seems impossible, because it is either there or not. Putting trust on the agenda seems like the first step towards its evaporation.

Having said that, I would like to put forward three models. All three engage with the question of who gets to decide in the matter of 'good' and 'bad' information – or for that matter, of 'expertise' and 'non expertise'. In all three cases, the participants in the debate realize that it is not power that is at stake, but rather trust and democratic authority. In his book *Trust in Numbers* Ted Porter is presenting an analytical distinction between 'disciplinary objectivity' and 'mechanical objectivity'.[59] I would like to suggest that they may in fact be concepts describing a temporal development. The first concept refers to consensus among professional experts; it is the kind of consensus that is beyond doubt because laypeople put trust in experts, granting them a social mandate to engage in whatever activity they have been trained into. The explanations given by the gentleman-physician tend to be personal and clinical, providing reassurance and consolation to a bewildered public. Whenever consensus among experts cannot be reached or does not satisfy outsiders (or is even distrusted by them), there is a need for mechanical objectivity. This is formalized, bureaucratic knowledge with a seemingly self-evident character, meant to satisfy the general public. Modern evidence-based medicine may be considered a good example of mechanical objectivity, providing the scientific answer to a moral demand for impartiality and fairness in a democratic mass-society.

In *The Paradox of Scientific Authority* Wiebe Bijker, Roland Bal and Ruud Hendriks examine the way in which the Health Council – the highest advisory body in health issues in the Netherlands – makes an effort in giving authoritative advice to the minister of health, even in complex matters.[60] By introducing a spatial distinction between back-stage and front-stage, they show how the scientific advisory committees of the Health Council succeed in speaking with an authoritative voice, being fully aware that their advice has been constructed. Indeed, it is the negotiated consensus reached between experts in the field and the relevant audiences of the Health Council. Although the dichotomy between front-stage and back-stage may be rather artificial, upholding the paradox seems an important attempt to solve the tension between the inherent uncertainty of scientific knowledge and the societal need for scientific authority.

A third suggestion with the most radical implications is made by Roger Pielke, in *The Honest Broker*.[61] Whereas in the model proposed by Bijker, scientific experts are the most dominant ones to decide on the state of scientific expertise and about what should be done in the public domain, Pielke is including all of us. The defining characteristic of the honest broker of policy alternatives is that it engages in decision-making by clarifying and, at times, expanding the scope of choice available to decision-makers, in a way that allows the decision-maker to reduce choice based on his or her own preferences and values. In this model, the scientific expert is supplying scenarios, leaving it up to decision-makers to decide on the best scenario in a given situation. In a way, one could say the tension between disciplinary and mechanical objectivity as well as between back-stage and front-stage has evaporated in this model. The problem with it may be that it is not fully clear how to operationalize or even institutionalize it, to prevent science from becoming politicized and paralyzed.

This may be the greatest paradox of all: how to politically secure democratic consensus in our knowledge society regarding spiralling health care costs, distributive justice in health care and accessibility to the healer of one's choice. We may wonder if things haven't gotten out of hand, now that the corrosion of public trust in scientific expertise is a fact – as the vaccination crisis of 2009 seems to suggest. Because no society can afford the luxury of living without moral guidelines and social cohesion, it is time to reflect on the foundations of our health care system.[62]

NOTES

Introduction: Performing Expertise

1. For a detailed analysis of A. G. Doiarenko's career as a scientist and expert, see chapter 6 by Katja Bruisch in this volume.

2. T. Broman, 'The Semblance of Transparency: Expertise as a Social Good and an Ideology in Enlightened Societies', in R.E. Kohler and K.M. Olesko (eds), *Clio Meets Science. The Challenges of History*, (Osiris, 27, 2012), pp. 188–208, on pp. 188–93.

3. Some examples of edited collections in the history of expertise in the last decade: B. Ziemann, K. Brückweh, D. Schumann and R. Wetzell (eds), *Engineering Society. The Role of the Human and Social Sciences in Modern Societies, 1880–1980* (Houndmills: Palgrave 2012); M. Kohlrausch, K. Steffen and S. Wiederkehr (eds), *Expert Cultures in Central Eastern Europe. The Internationalization of Knowledge and the Transformation of Nation States Since World War I* (Osnabrück: Fibre, 2010); C. Rabier (ed.), *Fields of Expertise: A Comparative History of Expert Procedures in Paris and London, 1600 to Present* (Newcastle: Cambridge Scholars Publishing, 2007); F. Van Lunteren and R. Vermij (eds), *De opmars van deskundigen. Souffleurs van de samenleving* (Experts on the March. Prompters of Society) (Amsterdam: Amsterdam University Press, 2002).

4. For a general overview: K. Olesko, 'Historiography of Science', in J.L. Heilbron (ed.), *The Oxford Companion to the History of Modern Science* (Oxford: Oxford University Press, 2003), pp. 366–70. On the most recent trends in the history of science: R.E. Kohler and K.M. Olesko (eds), *Clio meets Science. The Challenges of History*, (Osiris, 27, 2012).

5. L. Raphael, 'Die Verwissenschaftung des Sozialen als methodische und konzeptuelle Herausforderung für eine Sozielgeschichte des 20. Jahrhunderts', *Geschichte und Gesellschaft*, 22:2 (1996), pp. 165–93.

6. H. M. Collins and R. Evans, 'The Third Wave of Science Studies: Studies of Expertise and Experience', *Social Studies of Science*, 32 (2002), pp. 235–96; H. M. Collins and R. Evans, *Rethinking Expertise* (Chicago, IL: University of Chicago Press, 2007); N. Stehr and R. Grundmann, *Experts. The Knowledge and Power of Expertise* (London and New York, NY: Routledge, 2011); E. Kurz-Milcke and G. Gigerenzer (eds), *Experts in Science and Society* (New York, NY: Kluwer Academic/Plenum Publishers, 2004); K.A. Ericsson et al. (eds), *The Cambridge Handbook of Expertise and Expert Performance* (Cambridge: Cambridge University Press, 2006).

7. E. Goffman, *The Presentation of Self in Everyday Life* (Edinburgh: University of Edinburgh, 1956).

8. S. Hilgartner, *Science on Stage: Expert Advice as Public Drama* (Stanford, CA: Stanford

University Press, 2000).

9. W. Bijker, R. Bal and R. Hendriks, *The Paradox of Scientific Authority: The Role of Scientific Advice in Democracies* (Cambridge, MA: MIT Press, 2009). On these models, see also the contribution by Frank Huisman in this volume. Roger Pielke presented the expert as an honest broker, who mediated between different audiences and whose attainment of expert authority clearly depends on this mediation: R.A. Pielke, *The Honest Broker: Making Sense of Science in Policy and Politics* (Cambridge: Cambridge University Press, 2007). On the mediating role of experts, see also: M. Callon and A. Rip, 'Humains, non-humains: morale d'une coexistence', in J. Theys and B. Kalaora (eds), *La Terre outragée: les experts sont formels* (Paris: Autrement, 1992), pp. 161–82. The work of Sheila Jasanoff has been central to this research tradition: S. Jasanoff, *The Fifth Branch. Science Advisers as Policy Makers* (Cambridge, MA: Harvard University Press, 1990).

10. See e.g. M. Callon (ed.), *The Laws of the Market* (Oxford: Blackwell Publishers 1998).

11. D. MacKenzie, *An Engine, Not A Camera. How Financial Models Shape Markets* (Cambridge, MA: MIT Press, 2006) pp. 1–36.

12. On the relation and growing separation between the history of science and science studies: L. Daston, 'Science Studies and the History of Science', *Critical Inquiry*, 35 (2009), pp. 778–815. Some of the historical reserves of the study of expertise in the science studies: Broman, 'The Semblance of Transparency', pp. 187–193.

13. T. Golan, *Laws of Man and Laws of Nature: A History of Scientific Expert Testimony* (Cambdrige, MA: Harvard University Press, 2004). For a continental example, see P. Godding, *L'évolution de l'expertise en tant que preuve judiciaire, de l'Antiquité au 21e siècle* (Brussels: Académie royale de Belgique, Classe des Lettres, 2011).

14. For an introduction into this field of study: E. H. Ash, 'Introduction. Expertise and the Early Modern State', in E. H. Ash (ed.), *Expertise. Practical Knowledge and the Early Modern State* (*Osiris*, 25, 2010), pp. 1–24, on p. 1–3, 18 and 24.

15. E. H. Ash, *Power, Knowledge, and Expertise in Elizabethan England* (Baltimore, MD: Johns Hopkins University Press, 2003).

16. A. Wakefield, 'Leibniz and the Wind Machines', in Ash, *Expertise*, pp. 171–88.

17. U. Klein and E. C. Spary (eds), *Materials and Expertise in Early Modern Europe: Between Market and Laboratory* (Chicago, IL: University of Chicago Press, 2010), p. 22.

18. See for example: J. E. McClellan III, *Science Reorganized. Scientific Societies in the Eighteenth Century* (New York, NY: Columbia University Press, 1985).

19. Klein and Spary, *Materials and Expertise*, p. 6.

20. U. Klein, 'Introduction: Artisanal-Scientific Experts in Eighteenth-century France and Germany', *Annals of Science*, 69:3 (2012), pp. 303–6.

21. R. Macleod (ed.), *Government and Expertise: Specialists, Administrators and Professionals, 1860–1919* (Cambridge: Cambridge University Press, 1988); J. Evetts, A. H. Mieg and U. Felt, 'Professionalization, Scientific Expertise, and Elitism: A Sociological Perspective', in Ericsson et al. (eds), *The Cambridge Handbook*, pp. 105–23.

22. See for example: M. Kohlrausch and H. Trischler, *Building Europe on Expertise: Innovators, Organizers, Networkers* (Basingstoke: Palgrave Macmillan, 2014), pp. 5–13; M. Kohlrausch, K. Steffen and S. Wiederkehr, 'Introduction', in Kohlrausch, Steffen and Wiederkehr (eds), *Expert Cultures in Central Eastern Europe*, pp. 9–10; L. Raphael, 'Embedding the Human and Social Sciences in Western Societies, 1880–1980: Reflections on Trends and Methods of Current Research', in Ziemann, Brückweh, Schumann and Wetzell (eds), *Engineering Society*, pp. 41–56, on p. 48–54. Martin Kohlrausch also discusses

the historical turning points in the history of expertise in his chapter in this volume.

23. D. Pestre, 'Regimes of Knowledge Production in Society. Towards a More Political and Social Reading', *Minerva*, 41 (2003), pp. 245–261. The development of public health forms an example of both increased supply and demand for scientific advice, see for example: D. Porter (ed.), *The History of Public Health and the Modern State* (Amsterdam: Rodopi, 1994).

24. Broman, 'The Semblance of Transparency'.

25. Ash, 'Introduction', p. 8.

26. T. M. Porter, *Trust in Numbers: The Pursuit of Objectivity in Science and Public Life* (Princeton, NJ: Princeton University Press, 1995); G. Gooday, '"Vague and Artificial". The Historically Elusive Distinction between Pure and Applied Science', *Isis*, 102:3 (2012), pp. 546–54.

27. Van Lunteren and Vermij, *De opmars van deskundigen*, pp. 9–10; See also: T. L. Haskell, *The Authority of Experts: Studies in History and Theory* (Bloomington, IN: Indiana University Press, 1984).

28. On the credibility of experts, see: G. Gooday, 'Liars, Experts and Authorities', *History of Science*, 46 (2008), pp. 431–56. On the matter of scientific expertise and financing: C. Mukerji, *A Fragile Power: Scientists and the State* (Princeton, NJ: Princeton University Press, 1989), pp. 4–12.

29. On expertise and social reform: C. Leonards and N. Randeraad, 'Transnational Experts and Social Reform, 1840–1880', *International Review of Social History*, 55:2 (2010), pp. 215–39; I. de Haan, J. ten Have, J. Kennedy and P. J. Knegtmans (eds), *Het eenzame gelijk. Hervormers tussen droom en daad* (Solitary in their Right. Reformers between Word and Deed) (Amsterdam: Boom, 2009). An example from the history of technology: J. K. Alexander, *The Mantra of Efficiency: From Waterwheel to Social Control* (Baltimore, MD: The Johns Hopkins University Press, 2008).

30. See also: S. Fisch and W. Rudloff (eds), *Experten und Politik. Wissenschaftliche Politikberatung in Geschichtlicher Perspektive* (Berlin: Duncker & Humblot, 2004).

31. Kohlrausch and Trischler, *Building Europe on Expertise*, pp. 3–6.

32. Kohlrausch and Trischler, *Building Europe on Expertise*, p. 5.

33. See for example: S. Turner, 'What is the Problem with Experts?', *Social Studies of Science*, 31 (2001), pp. 123–49; S. Turner, *Liberal Democracy 3.0: Civil Society in an Age of Experts* (London: Sage, 2003); L. Dumoulin, S. Labranche, C. Robert and P. Warin (eds), *Le recours aux experts. Raisons et usages politiques* (Grenobles: Presses universitaires de Grenoble, 2005).

1 Ethnicity, Expertise and Authority: The Cases of Lewis Howard Latimer, William Preece and John Tyndall

1. 'Oral Telegraphy', *Times*, 19 September 1877 p. 4.

2. See J. Vandendriessche, E. Peeters and K. Wils, 'Introduction: Performing Expertise' in this volume, discussing E. Ash, *Power, Knowledge, and Expertise in Elizabethan England* (Baltimore, MD: Johns Hopkins University Press, 2003).

3. G. Gooday, 'Fear, Shunning, and Valuelessness: Controversy over the Use of "Cambridge" Mathematics in Late Victorian Electro-Technology', in D. Kaiser (ed.), *Pedagogy and the Practice of Science: Historical and Contemporary Perspectives* (Cambridge, MA: MIT Press, 2005), pp. 111–49.

4. Although Bell took classes at the University of London before he and his family migrated to Canada in 1870, his primary training from his father and grandfather was as a teacher of deaf people. See R. V. Bruce, *Bell: Alexander Graham Bell and the Conquest of Solitude* (Ithaca, NY: University of Cornell Press, 1976).

5. 'Oral Telegraphy', *The Times*, 19 September 1877.

6. See further discussion in S. Arapostathis and G. Gooday, *Patently Contestable: Historical Trials of Electricity, Identity, and Inventorship* (Cambridge, MA: MIT Press, 2013), pp. 22–3.

7. W. B. Carlson, *Tesla: Inventor of the Electrical Age* (Princeton, NJ: Princeton University Press, 2013).

8. G. Gooday, 'Liars, Experts and Authorities', *History of Science*, 46 (2008), pp. 431–56.

9. T. Golan, *Laws of Men, Laws of Nature: The History of Scientific Expert Testimony in England and America* (Cambridge, MA: Harvard University Press, 2007).

10. Gooday, 'Liars, Experts and Authorities', p. 431.

11. G. Gooday, *Domesticating Electricity: Technology, Gender and Uncertainty, 1880–1914* (London: Pickering and Chatto, 2008). The only exception is Hertha Ayrton addressing the Institution of Electrical Engineers in 1899 on the arc light. See J. Mason, 'Hertha Ayrton', in N. Byers and G. Williams (eds), *Out Of The Shadows: Contributions of 20th Century Women to Physics* (Cambridge: Cambridge University Press, 2006), pp. 15–25.

12. P. Nahin, *Heaviside: The Life, Work, and Times of an Electrical Genius of the Victorian Age*, rev. edn (Baltimore, MD: Johns Hopkins Press, 2002).

13. We cannot otherwise make sense of why only certain topics have elicited a huge popularization in literature (e.g. electricity and evolution) and others have not (e.g. magnetism and mining).

14. This account is based in part on chapter 3: 'Lewis H. Latimer and the Politics of Technological Assimilitationism' in R. Fouché, *Black Inventors in the Age of Segregation: Granville T. Woods, Lewis Howard Latimer and Shelby J. Davidson* (Baltimore, MD: Johns Hopkins University Press, 2003), pp. 82–133. For broader context, see B. Sinclair (ed.), *Technology and the African-American Experience: Needs and Opportunities for Study* (Cambridge, MA: MIT Press, 2004). My thanks to Paul Israel for directing my attention to the invaluable public summary of Latimer's career in the Exhibition Catalogue by J. M. Schneider and B. Singer, 'Blueprint for Change: The Life and Times of Lewis H. Latimer', at https://edison.rutgers.edu/latimer/catalog.htm [accessed 15 February 2014].

15. 'Electrical Recollections by L.H.Latimer', (unpaginated) in Lewis Howard Latimer papers, Queens Borough Public Library, New York, NY, Biographical material – 376/25.

16. Logbook (unpaginated) in Lewis Howard Latimer papers, Queens Borough Public Library, New York, NY, Biographical material – 376/25.

17. See, for example, the illustration of Latimer and team at http://edison.rutgers.edu/latimer/gelegal.htm [accessed 15 February 2014].

18. Report by attorney Sherburne B. Eaton to the Patent Litigation Committee, fa0631, Patents, Edison Papers, Rutgers University, Document File Series – 1891: (D-91– 42).

19. In 1924 ill-health led to Latimer's retirement from practical work: Lewis H Latimer papers, Queens Borough Public Library, New York, NY, Biographical material – 376/25, p. 17.

20. Edison Pioneers Membership Certificate (1921) at http://edison.rutgers.edu/latimer/latpio.htm [accessed 15 February 2014].

21. Fouché, *Black Inventors in the Age of Segregation*, pp.122–33.

22. See Arapostathis and Gooday, *Patently Contestable: Historical Trials of Electricity*,

Identity, and Inventorship, pp. 59–85.

23. Gooday, *Domesticating Electricity*, chapters 2 and 4.
24. E. C. Baker, *Sir William Preece, F.R.S: Victorian Engineer Extraordinary* (London: Hutchinson, 1976).
25. Gooday, *Domesticating Electricity*, chapters 2 and 4.
26. S. Arapostathis, 'Electrical Innovations, Authority and Consulting Expertise in Late Victorian Britain', *Notes and Records of the Royal Society*, 67:1 (March 2013), pp. 59–76.
27. See for example, 'Sir William Henry Preece, A Working Portrait', 4 December 1892, BT Archives, TCE 361/ARC 704.
28. E. Bruton, 'Beyond Marconi: The Roles of the Admiralty, the Post Office, and the Institution of Electrical Engineers in the Invention and Development of Wireless Communication up to 1908' (PhD dissertation, University of Leeds, 2012).
29. O. Heaviside, *Electrical Papers. Volume 1* (London: Macmillan, 1892), pp. v-vii. This is not quite in the mould of the self-absorbed hermit as some biographers have caricatured him: D. W. Jordan, 'The Adoption of Self-Induction by Telephony', *Annals of Science*, 39 (1982), pp. 433–61. For detailed discussion see Arapostathis and Gooday, *Patently Contestable*, pp. 106–10.
30. 'The Eminent Scienticulist' was one of Heaviside's many nicknames for Preece; others included 'the Bouncer', 'Mr. Prigs' and 'Taffy'; B. Hunt, *The Maxwellians* (Ithaca, NY: Cornell University Press, 1991), pp. 140–2.
31. See W.C. Baker, *Sir William Preece, Victorian Engineer Extraordinary* (Hutchinson, 1976), pp. 342–52.
32. See the illustration of John Tyndall Lecturing at the Royal Institution, from *Illustrated London News*, 14 May 1870 shown in J. Howard, '"Physics and fashion": John Tyndall and his Audiences in Mid-Victorian Britain', *Studies in History and Philosophy of Science*, 35 (2004), pp. 729–58.
33. See B. Lightman, 'Scientists as Materialists in the Periodical Press: Tyndall's Belfast Address' in G. Cantor and S. Shuttleworth (eds), *Science Serialized: Representations of the Sciences in Nineteenth-Century Periodicals* (Cambridge, MA: MIT Press, 2004), pp. 199–238. Tyndall's provocative remarks about religion prompted a complaint in 1877 from the President of the Royal Institution, the sixth Duke of Northumberland. Tyndall defended himself by differentiating between his professorial prerogatives at the RI and his private proclivities when writing from the Athenaeum Club: the 'clearly understood law of the Royal Institution' was that neither religion nor politics could be introduced into his lectures there; but when writing or speaking outside those walls he claimed 'intellectual freedom'. See John Tyndall to William Spottiswoode, 11 November 1877, Tyndall papers, Royal Institution, MS JT/TYP/1264 – 1265.
34. Spottiswoode to Tyndall, 29 December 1874, Tyndall papers, Royal Institution.
35. Tyndall to Galton [cc: Spottiswoode], 30 December 1877, Tyndall papers, Royal Institution, MS JT/TYP/1266.
36. O. Lodge, *Past Years* (London: Hodder & Stoughton, 1931), pp. 65–6.
37. S. P. Thompson to his mother, spring 1876 (cited in H. Thompson and J. Thompson, *Silvanus Phillips Thompson: His Life and Letters* (London: T. Fisher Unwin, 1920), p. 21.)
38. Thompson and Thompson, *Silvanus Phillips Thompson*, p. 41.
39. Thompson and Thompson, *Silvanus Phillips Thompson*, p. 41. Huxley himself noted after Tyndall's death that he was one of the 'very few orators whom I have heard to whom I could not choose but listen.' T. H. Huxley, 'Professor Tyndall', *Nineteenth Century*, 35 (1894), p. 5.

40. Thompson and Thompson, *Silvanus Phillips Thompson*, p. 24.
41. Jane Thompson reported Preece's post-lecture comment. 'When you and I, Thompson, come to that stage, I hope some friend will be kind enough to prevent our making an exhibition of ourselves'. Thompson and Thompson, *Silvanus Phillips Thompson*, p. 157.
42. 'Science Notes', *Academy*, 722 (1886: Mar. 6), p. 171.
43. Initially it was reported in the press that a comeback might be possible, 'Science Gossip', *The Athenaeum* (6 November 1886), p. 605. But it was clear by the following spring that Tyndall had been obliged by his failing health to resign from his position at the Royal Institution, and Lord Rayleigh was appointed as his successor. 'Science Notes', *Academy*, 780 (16 April 1887), p. 278. See also the tribute: 'Professor Tyndall', *Saturday Review*, 63 (2 April 1887), p. 468.
44. S. P. Thompson, 'Obituary – Professor Tyndall', *The Electrician*, 8 December 1893, pp. 141–2.
45. O. Lodge, 'John Tyndall', *Encyclopaedia Britannica*, 10th edn (Edinburgh & London: Adam & Charles Black, 1902–3), pp. 517–8.
46. Early versions of this paper were also presented at the University of Valencia and the SHOT 2012 meeting in Copenhagen. Quotations from the John Tyndall correspondence at the Royal Institution arise from a British Academy-funded project in 2008 to transcribe the letters of Tyndall to and from James Joule, Lyon Playfair, William Spottiswoode and William Thomson, and from participation subsequently in the International National Science Foundation-funded Tyndall Correspondence project. I thank Anne Locker at the Archives of the Institution of Electrical Engineers for access to the William Preece papers, and for staff at the Borough public library of Queens, New York, for the opportunity to see the papers of Lewis Howard Latimer.

2 Arbiters of Science: Expertise in Public Health in Ninteenth-Century Belgian Medical Societies

1. 'Séance extraordinaire du 26 Janvier 1864', *Bulletin de la Société de Médecine de Gand* (hereafter *BSMG*), 31 (1864), p. 32; A. Burggraeve, *Question sociale. Amélioration de la vie domestique de la classe ouvrière* (Ghent: De Busscher, 1864).
2. For example: A. Burggraeve, *Histologie ou Anatomie de texture* (Ghent: Annoot-Brackman, 1843); A. Burggraeve, *Mémoire sur les appareils ouatés* (Ghent: Gyselinck, 1850).
3. F. Adriaensen, 'Dr. Adolphe Burggraeve: arbeider als patiënt, stad als panoptikum', *Tijdschrift voor Geschiedenis van Techniek en Industriële Cultuur*, 9 (1991), pp. 4–23.
4. K. Velle, 'De centrale gezondheidsadministratie in België voor de oprichting van het eerste ministerie van volksgezondheid (1849–1936)', *Belgisch Tijdschrift voor Nieuwste Geschiedenis*, 11 (1990), pp. 162–210.
5. K. Velle, *De nieuwe biechtvaders. De sociale geschiedenis van de arts in België* (Leuven: Kritak, 1991), p. 78.
6. For a recent discussion of the historiography of public health: C. Hamlin, *Cholera: The Biography* (Oxford: Oxford University Press, 2009). On the interaction between public health experts and the state: P. Baldwin, *Contagion and the State in Europe, 1830–1930* (Cambridge: Cambridge University Press, 1999); D. Porter (ed.), *The History of Public Health and the Modern State* (Amsterdam: Rodopi, 1994).
7. The argument of the mutually reinforcing relation between experts and the state was first made by Oliver MacDonagh in his seminal article of 1958: O. MacDonagh, 'The

Nineteenth Century Revolution in Government: A Reappraisal', *Historical Journal*, 1:1 (1958), pp. 52–67. For a discussion of MacDonagh's thesis and its influence on historiography, see: R. Macleod (ed.), *Government and Expertise: Specialists, Administrators and Professionals, 1860–1919* (Cambridge: Cambridge University Press, 1988), pp. 3–11.

8. F. Chevallier, *Le Paris moderne: Histoire des politiques d'hygiène, 1855–1898*. (Rennes: Presses universitaires de Rennes, 2010); D.S. Barnes, *The Great Stink of Paris and the Nineteenth-Century Struggle against Filth and Germs* (Baltimore, MD: Johns Hopkins University Press, 2006).

9. For a performative analysis of medical societies: M. Brown, *Performing Medicine: Medical Culture and Identity in Provincial England, c.1760–1850* (Manchester: Manchester University Press, 2011).

10. A. Despy-Meyer, 'Instellingen en netwerken', in R. Halleux, J. Vandersmissen, A. Despy-Meyer, and G. Vanpaemel (eds), *Geschiedenis van de wetenschappen in België 1815–2000* (Brussels: Dexia bank, 2001), pp. 71–90, on pp. 87–89.

11. 'Liste des membres', *BSMG*, 6 (1840), pp. 5–17.

12. 'Liste des Membres titulaires de la Société', in *Société Royale des Sciences Médicales et Naturelles de Bruxelles. Volume jubilaire publié à l'occasion du centenaire de la Société Royale des sciences médicales et naturelles de Bruxelles* (Brussels: Lamertin, 1922), pp. 57–61.

13. On the development of public health institutions in Belgium: E. Bruyneel, *De Hoge Gezondheidsraad (1849–2009). Schakel tussen wetenschap en volksgezondheid* (Leuven: Peeters, 2009), pp. 17–33.

14. B. C. Ingels, 'Le professeur Alexis-César Lados, Président honoraire de la Société de Médecine de Gand', *BSMG*, 47 (1880), pp. 56–64, on p. 58, 'Des travaux aussi remarquables ... finirent par attirer sur lui l'attention du Gouvernement'.

15. 'Le professeur Alexis-César Lados, Président honoraire de la Société de Médecine de Gand', pp. 58–60.

16. 'Sociétés littéraires et scientifiques', 1851, Ghent City Archive, T 521/357.

17. On the mechanism of reciprocal legitimization between experts and the state: E. Ash, 'Introduction: Expertise and the Early Modern State', *Osiris*, 25 (2010), pp. 1–24; On Belgian social policy: E. Witte, J. Craeybeckx, and A. Meynen, *Politieke geschiedenis van België van 1830 tot heden* (Antwerp: Standaard, 2010), pp. 71–77.

18. D. Mareska and J. Heyman, *Enquête sur le travail et la condition physique et morale des ouvriers employés dans les manufactures de coton, à Gand* (Ghent: Gyselinck, 1845).

19. 'Séance du 8 Juillet 1845', *BSMG*, 11 (1845), pp. 161–2, on p. 162, 'couvert d'unanimes applaudissements'.

20. E. Witte et al., *Nieuwe geschiedenis van België, 1830–1905* (Tielt: Lannoo, 2005); C. Verbruggen, *De stank bederft onze eetwaren. De reacties op industriële milieuhinder in het 19de-eeuwse Gent* (Ghent: Academia Press, 2002), pp. 12–13; G. Deneckere, *Katoenoproer van Gent in 1839: collectieve actie en sociale geschiedenis* (Nijmegen: SUN, 1998); R. Mantels, *Gent. Een geschiedenis van universiteit en stad, 1817–1940* (Ghent: UGent Memorie/ Mercatorfonds, 2014), 14–15.

21. Hereafter francs to be written F.

22. 'Séance ordinaire du mois de Décembre', *BSMG*, 28 (1861), pp. 414–16, on p. 414, 'Quelles sont les notions hygiéniques applicables aux établissements destinés à l'instruction de la jeunesse'.

23. 'Séance ordinaire du 11 Février 1862', *BSMG*, 29 (1862), pp. 39–40, on p. 39: 'Décrire les maladies qui peuvent tirer leur origine de l'exercice des industries linière et cotonnière'.

24. E. Ash, *Power, Knowledge, and Expertise in Elizabethan England* (Baltimore, MD:

Johns Hopkins University Press, 2003); C. Hannaway, 'The Société Royale de Médecine and Epidemics in the Ancien Régime', *Bulletin of the History of Medicine*, 46 (1972), pp. 257–73.

25. J.F.J. Dieudonné, *Mémoire sur la condition des classes ouvrières et sur le travail des enfants* (Brussels: Lesigne, 1846).

26. 'Discussion sur la note qui précède', *BSMG*, 28 (1861), pp. 245–50, on p. 246, 'une forte prime'.

27. In Ghent, both the mayor, Charles de Kerckhove, and the secretary of internal affairs, Alphonse Van den Peereenboom, received such titles in 1867: 'Séance ordinaire du mois de Mars 1867', *BSMG*, 34 (1867), pp. 33–5, on p. 33. In Brussels, mayor Jules Anspach became an honorary member in 1872: 'Bulletin de la séance du 5 août 1872', *Journal de médecine, de chirurgie et de pharmacologie* (hereafter *Journal de médecine*), 55 (1872), pp. 177–80, on p. 178.

28. 'Visite à la Famille royale', *BSMG*, 27 (1860), pp. 273–74, on p. 274: 'Continuez, Messieurs, de concert avec mon Gouvernement à travailler à l'amélioration de l'état sanitaire de ces contrées'.

29. 'Séance ordinaire du 2 Juin 1863', *BSMG*, 30 (1863), pp. 205–7, on p. 205.

30. Velle, *De nieuwe biechtvaders*, pp. 105–114.

31. See for example the two chapters on medical education in Belgium in: P. Dhondt, *Un double compromis: enjeux et débats relatifs à l'enseignement universitaire en Belgique au XIXe siècle* (Ghent: Academia Press, 2011), pp. 225–43 and 361–84.

32. For a general overview of popularization of science: D.M. Knight, 'Scientists and Their Public: Popularization of Science in the Nineteenth Century', in M.J. Nye (ed.), *The Modern Physical and Mathematical Sciences* (Cambridge History of Science, vol. 5) (Cambridge: Cambridge University Press, 2002), pp. 72–90.

33. A. Burggraeve, *Art de prolonger la vie* (Brussels: Office de Publicité, 1868); A. Burggraeve, *Médecine populaire. De l'homme physique* (Ghent: Hoste, 1853–1854); A. Burggraeve, *Le livre de tout le monde sur la santé: notions de physiologie et d'hygiène* (Paris: Didier, 1863); A. Burrgraeve, *Hygiène populaire. Longévité humaine, ou art de prolonger la vie* (Brussels: Lesigne, 1876).

34. Charles Van Bambeke, who was later to be appointed professor, joined the society as assistant-surgeon: 'Séance ordinaire du 7 Février 1860', *BSMG*, 27 (1860), pp. 53–7, on p. 53; Van Duyse joined as *préparateur* in the course of pathological anatomy: 'Séance ordinaire du 2 Mars 1880', *BSMG*, 47 (1880), pp. 113–15, on p. 115.

35. On the profile of the members of medical societies: S.C. Lawrence, '"Desirous of Improvements in Medicine". Pupils and Practitioners in the Medical Societies at Guy's and St. Bartholomew's Hospitals, 1795–1815', *Bulletin of the History of Medicine*, 59 (1985), pp. 89–104.

36. 'J.C. Coppée, 'Rapport sur le mémoire intitulé: Quelques considérations sur la question de l'hygiène des hôpitaux, par M. le docteur Frédericq', *BSMG*, 30 (1863), pp. 161–69, on p. 161, 'tout travail traitant de l'hygiène des hôpitaux, sera accueillie par nous avec faveur; nous n'y mettons qu'une réserve, c'est que nous nous souviendrons aussi que le désir de bien faire peut parfois entraîner à l'exagération'.

37. Coppée, 'Rapport sur le mémoire intitulé', p. 164, 'une donnée très-générale'.

38. L. Frédericq, *Handboek van gezondheidsleer voor alle standen* (Ghent: Rogghé, 1867); L. Frédericq, *Hygiène populaire* (Ghent: Hoste, 1875); L. Frédericq, *Lichaamsongelukken hulpmiddelen vóór de aankomst van den geneesheer* (Ghent: Vuylsteke, 1882).

39. Frédericq, *Handboek van gezondheidsleer*, p. 261; See also: Velle, *De nieuwe*

biechtvaders, p. 295.

40. 'Séance ordinaire du 2 Juin 1868', *BSMG*, 35 (1868), pp. 237–40, on p. 239: 'ce travail n'est que la reproduction incomplète, mais dans certaines passages, servilement littérale d'une petite brochure ... par Jac. Dycer'.

41. 'Séance ordinaire du 3 Septembre 1867', *BSMG*, 34 (1867), pp. 289–91, on p. 290.

42. S. Van Damme, 'Foreword. Expertise in Capital Cities', in C. Rabier (ed.), *Fields of Expertise: A Comparative History of Expert Procedures in Paris and London, 1600 to Present* (Newcastle: Cambridge Scholars Publishing, 2007), pp. xiv–xv.

43. 'Bulletin de la séance du 8 Janvier 1872', *Journal de médecine*, 54 (1872), pp. 78–82, on pp. 79–80: 'Ce mémoire ... appartient plus au domaine administratif qu'à la science proprement dite. Cependant, l'hygiène publique, qui en fait l'objet, a une importance trop marquée pour que la Société ne s'intéresse pas à toutes les mesures tendant à en propager l'application et, surtout, l'organization'.

44. On Janssens, see: K. Velle, 'Janssens, Eugène Dorothé', *Nationaal Biografisch Woordenboek*, vol. 10, pp. 302–7.

45. 'Bulletin de la séance du 4 décembre', *Journal de médecine*, 75 (1882), pp. 580–7, on p. 584: 'une véritable épidémie de bonne santé'.

46. 'Bulletin de la séance du 7 janvier 1878', *Journal de Médecine*, 66 (1878), pp. 84–91, on p. 85.

47. 'Bulletin de la séance du 9 janvier 1860', *Journal de médecine*, 30 (1860), pp. 182–92, on pp. 189–90.

48. J.-L. Delaet, 'Rationalisme et progressisme au Pays de Charleroi. Biographie du Docteur Hubert Boëns (1825–98)', *Documents et rapport de la Société royale d'archéologie et de paléontologie de Charleroi*, 60 (1986–88), pp. 156–67; K. Wils, *De omweg van de wetenschap. Het positivisme en Belgische en Nederlandse intellectuele cultuur 1845–1914* (Amsterdam: Amsterdam University Press, 2005), p. 173; G. Leboucq, 'Hubert Boëns', *Biographie Nationale*, vol. 29, p. 309.

49. 'Bulletin de la séance du 5 mars 1860', *Journal de médecine*, 30 (1860), pp. 412–8, on pp. 412–4.

50. 'Bulletin de la séance du 5 mars 1860', on p. 413: 'de trop écrire et de ne pas assez réfléchir'.

51. 'Bulletin de la séance du 2 avril 1860', *Journal de médecine*, 30 (1860), pp. 515–32, on p. 532: 'En résumé, je pense que le travail de M. Loneux peut être publié dans notre journal, parce que la discussion ... prouvera suffisamment à nos lecteurs que nous n'adaptons pas sans réserve les opinions de l'honorable médecin d'Hérenthals'.

52. 'Bulletin de la séance du 1er juin 1874', *Journal de médecine*, 58 (1874), pp. 565–71, on pp. 570–1: 'il ne s'agira plus de limiter notre intervention à un rôle exclusivement scientifique, nous devons exercer une influence politique, vulgariser et défendre par la voie de la Presse le nouveau système que nous voulons faire prévaloir; marches auprès des divers pouvoirs de la nation et provoquer un pétitionnement général'.

53. V. Deneffe, 'Projet d'inhumation par l'incrustation des corps dans des pierres artificielles, par Louis Cruls, ingénieur du gouvernement brésilien', *BSMG*, 43 (1876), pp. 124–8, on p. 128, 'j'ai peine à croire que le projet de M. Cruls [...] aie grande chance d'être adopté par les administrations'.

54. Evert Peeters and Kaat Wils have shown that the question of state intervention in public health divided liberal politicians and physicians alike and required the balancing of the liberal ideals of individual freedom and rational progress, of which the state was seen as the main vehicle: E. Peeters and K. Wils, 'Ambivalences of Liberal Health Policy: *Lebensreform* and Self-Help Medicine in Belgium, 1890–1914', in F. Huisman and H. Oosterhuis (eds), *Health and Citizenship* (London: Pickering and Chatto,

2014), pp. 101–17, on pp. 102–7.

55. Wils, *De omweg van de wetenschap*, pp. 170–4. More generally on the political engagement of Belgian physicians: Velle, *De nieuwe biechtvaders*, pp. 177–185; A. Morelli,
'Les médecins parlementaires belges (19e-20e siècles)', in *L'engagement social et politique
des médecins. Belgique et Canada, XIXe et XXe siècles. Actes du colloque tenu à l'U.L.B.
les 5 et 6 février 1993* (Brussels: Institut Emile Vandervelde, 1993), pp. 9–18.

56. H. Leboucq, 'Crocq (*Jean*)', *Biographie Nationale*, vol. 30, pp. 301–3; 'Crocq, *Jean*, Joseph', in J.-L. De Paepe and C. Raindorf-Gérard (eds), *Le Parlement belge 1831–1894.
Données biographiques* (Brussels: Académie royale de Belgique, 1996), p. 81; 'Pigeolet,
Arsène, Victor, Auguste', in De Paepe and Raindorf-Gérard (eds), *Le Parlement belge*,
pp. 462–3, on p. 462.

57. 'Bulletin de la séance du 1er août 1870', *Journal de médecine*, 51 (1870), pp. 165–71, on
pp. 166–7.

58. 'Bulletin de la séance du 7 décembre 1868', *Journal de médecine*, 47 (1868), pp. 568–77, on
p. 577: 'La liberté, rien que la liberté, voilà nos principes ... toucher à la liberté de la famille,
c'est une question qui ne rencontrera pas, dans cette enceinte, beaucoup de partisans'.

59. 'Bulletin de la séance du 5 octobre 1868', *Journal de médecine*, 47 (1868), pp. 378–88,
on pp. 386–8.

60. 'Bulletin de la séance du 5 novembre 1860', *Journal de médecine*, 31 (1860), pp. 616–37,
on p. 618: 'l'un des rouages de la grande machine'.

61. Velle, *De nieuwe biechtvaders*, pp. 266–72.

62. 'Bulletin de la séance du 5 octobre 1874', *Journal de médecine*, 59 (1874), pp. 365–71,
on p. 366; 'Bulletin de la séance du 6 mars 1876', *Journal de médecine*, 62 (1876), pp.
276–84, on pp. 282–4.

63. 'Bulletin de la séance du 5 janvier 1880', *Journal de médecine*, 70 (1880), pp. 88–105, on
pp. 89–92.

64. On the problems of 'disinterested expertise': N. Stehr, and R. Grundmann, *Experts.
The Knowledge and Power of Expertise* (London and New York, NY: Routledge, 2011);
T.M. Porter, *Trust in Numbers: The Pursuit of Objectivity in Science and Public Life*
(Princeton, NJ: Princeton University Press, 1995). More generally on the matter of
scientific expertise and financing: C. Mukerji, *A Fragile Power: Scientists and the State*
(Princeton, NJ: Princeton University Press, 1989), pp. 4–12.

65. I use the term 'authority' here following Graeme Gooday's discussion of authorities as
the social figures 'to which the laity could turn in its quest for unbiased impartial wis-
dom ... [and who did not] have any direct financial interests in matters of which they
pronounced': G. Gooday, 'Liars, Experts and Authorities', *History of Science*, 46 (2008),
pp. 431–56, on p. 432. For a further discussion of experts vis-à-vis authorities, see also
the contribution by Graeme Gooday in this volume.

66. 'Séance ordinaire du 7 Mai 1867', *BSMG*, 34 (1867), pp. 129–31, on p. 131: 'que l'état
sanitaire de la ville de Gand est très satisfaisant et que rien ne justifie sous ce rapport les
bruits alarmants que la légèreté ou la frayeur a fait courir'.

67. 'Bulletin de la séance du 2 octobre 1882', *Journal de médecine*, 75 (1882), pp. 397–400,
on p. 400.

68. 'Bulletin de la séance du 5 avril 1880', *Journal de médecine*, 70 (1880), pp. 412–29, on
p. 426: 'Je crois que notre société, où l'on trouve des hommes autorisés, tant au point de
vue scientifique qu'au point de vue pratique, est parfaitement en situation pour élucider
cette question et rassurer ainsi la confiance publique que des théoriciens imprudents
voudraient ébranler'.

69. 'Bulletin de la séance du 3 janvier 1876', *Journal de médecine*, 62 (1876), pp. 83–91, on p. 91: 'contributions à la désinfection des sépultures et à la crémation'.

70. In its annual report, the Medical Society of Brussels was said to be very pleased with the success of its sanitary bulletins: 'Bulletin de la séance du 6 juillet 1868', *Journal de médecine*, 47 (1868), pp. 68–85, on p. 71.

71. In 1880, the Medical Society of Ghent, for example, wrote to the town council after discussing a study on the transportation of persons stricken by contagious diseases: 'Séance ordinaire du 2 Mars 1880', *BSMG*, 47 (1880), p. 114.

72. 'Rapport sur les inconvénients que présentent les tuyaux en plomb pour l'aspiration de la bière', *BSMG*, 44 (1877), pp. 9–16, on p. 9.

73. I would like to thank Kaat Wils, Evert Peeters, Jacob Steere-Williams, Michael Brown and Darina Martykánová for their helpful suggestions on previous versions of this chapter.

3 Borderless Nature: Experts and the Internationalization of Nature Protection, 1890–1940

1. *The Proceedings of the 1st International Symposium of the Pan-European Ecological Network 'Nature Does Not Have Any Borders: Towards Transfrontier Ecological Networks'* (Paris: Council of Europe Publishing, 2000).

2. See, amongst others, J. Lovelock, *Gaia: A New Look at Life on Earth* (Oxford: Oxford University Press, 1979); R. Poole, *Earthrise: How Man First Saw The Earth* (New Haven, CT: Yale University Press, 2008); P. Blandin, 'Ecology and Biodiversity at the Beginning of the Twenty-First Century: Towards a New Paradigm?', in K. Jax and A. Schwarz (eds), *Ecology Revisited: Reflecting on Concepts, Advancing Science* (Dordrecht: Springer, 2011), pp. 205–14.

3. Some historians, in writing about this movement, have made a strong distinction between a tradition centreing on 'preservation' and one on 'conservation'. The exact difference between the two terms depends somewhat on the author, however. Furthermore, it seems that the historical reality shows not two opposing groups, but a great variety of partially overlapping traditions. I therefore believe, in line with Peder Anker, that the dichotomy is largely unhelpful and, in this article, only use the term conservation. P. Anker, *Imperial Ecology: Environmental Order in the British Empire, 1895–1945* (Cambridge, MA: Harvard University Press, 2001), p. 197.

4. The early twentieth century has overall received little attention of historians of the conservation movement. John McCormick's *Reclaiming Paradise* (1989), still the reference work on the issue, only deals with the period before 1940 in an introductory chapter. More recently Anna-Katharina Wöbse published a study on nature protection in the interwar years in the specific context of the League of Nations. Patricia van Schuylenbergh, then, published on the interaction between Belgian and Dutch actors in trans-border conservation initiatives. J. McCormick, *Reclaiming Paradise: The Global Environmental Movement* (London: Belhaven Press, 1989); P. van Schuylenbergh, 'Congo Nature Factory: wetenschappelijke netwerken en voorbeelden van Belgisch-Nederlandse uitwisselingen (1885–1940)', *Jaarboek voor Ecologische Geschiedenis, 2009*, 11 (2010), pp. 79–104; A.-K. Wöbse, *Weltnaturschutz: Umweltdiplomatie in Völkerbund und Vereinten Nationen 1920–1950* (Frankfurt: Campus Verlag, 2011).

5. See amongst others: M. Barrow, *Nature's Ghosts: Confronting Extinction from the Age of Jefferson to the Age of Ecology* (Chicago, IL and London: The University of Chicago

Press, 2009), pp. 135–67.

6. See: J. Sheail, *Nature in Trust: The History of Nature Conservation in Britain* (Glasgow: Blackie, 1976); W. Rollins, *A Greener Vision of Home: Cultural Politics and Environmental Reform in the German Heimatschutz Movement, 1904–1918* (Ann Arbor, MI: University of Michigan Press, 1997); C. Ford, 'Nature, Culture and Conservation in France and Her Colonies 1840–1940', *Past and Present*, 183 (2004), pp. 173–98.

7. On Conwentz: A. Milnik, *Hugo Conwentz– Klassiker des Naturschutzes. Sein Waldweg zum Naturschutz* (Kessel: Verlag Kessel Remagen, 2006).

8. H. Conwentz, *The Care of Natural Monuments with Special Reference to Great Britain and Germany* (Cambridge: Cambridge University Press, 1909), pp. 1–34.

9. On Heimatkunde: L. Nyhart, *Modern Nature: The Rise of the Biological Perspective in Germany* (Chicago, IL and London: The University of Chicago Press, 2009), pp. 163–4.

10. It was also from Humboldt that Conwentz borrowed the term 'natural monument'.

11. Yet, he did try to inventory birds' nests. Conwentz, *The Care of Natural Monuments*, pp. 36–44.

12. H. Conwentz, *Die Gefährdung der Naturdenkmäler und Vorschläge zu ihrer Erhaltung. Denkschrift, dem Herrn Minister der geistlichen, Unterrichts- und Medizinal-Angelegenheiten überreicht* (Berlin: Gebrüder Brontraeger, 1904), p. 207.

13. With regard to Germany, Switzerland and Belgium: Rollins, *A Greener Vision of Home*, pp. 119 and 160–1; P. Kupper, *Wildnis Schaffen: Eine transnationale Geschichte des Schweizerischen Nationalparks* (Bern: Haupt, 2012), pp. 184–90; A. Stynen, 'Vaderlandse weelde op de kaart gezet. Belgische botanici, wetenschappelijke ijver en vaderlandse motieven', *Bijdragen en mededelingen betreffende de geschiedenis der Nederlanden,* 121 (2006), pp. 680–710. Conwentz himself referred to the work of his colleague Eugenius Warming: Conwentz, *The Care of Natural Monuments*, p. 99. With regard to France: C. Flahault, 'La question forestière', *Bulletin de la Société de Botanique de France,* 38 (1891), pp. xxxix-xlv.

14. J. Schrijnen, *Nederlandsche Volkskunde* (Zutphen: W. J. Thieme en Cie, 1916), p. 343, quoted in W. Roenhorst, 'De natuurlijke natie. Monumentalisering en nationalisering van natuur en landschap in de vroege twintigste eeuw', *Bijdragen en mededelingen betreffende de geschiedenis der Nederlanden,* 121 (2006), pp. 727–52, on p. 742.

15. The history of scientific conferences as a place of scientific activity has received only little scholarly attention. A first overview is given in: W. Feuerhahn and P. Rabault-Feuerhahn (eds), *La Fabrique Internationale de la Science. Les Congrès Scientifiques de 1865 à 1945* (Paris: CNRS Éditions, 2010).

16. E. Stresemann, *Die Entwicklung der Ornithologie. Von Aristoteles bis zur Gegenwart* (Berlin: F.W. Peters, 1951), pp. 337–42; R. De Bont, 'Poetry and Precision. Johannes Thienemann, the Bird Observatory in Rossitten and Civic Ornithology', *Journal of the History of Biology*, 44 (2011), pp. 171–203.

17. S. Jansen, *Schädlinge: Geschichte eines wissenschaftlichen und politischen Konstrukts, 1840–1920* (Frankfurt: Frankfurt Verlag, 2003), pp. 60–3 and pp. 170–74.

18. The interventions were made by the Belgians, Edmond de Selys-Longchamps and Alfred Quinet, and the Swiss, Victor Fatio. See: É. Oustalet and J. de Claybrooke (eds), *IIIe Congrès Ornithologique International. Paris, 26–30 juin 1900. Compte rendu des séances* (Paris: Masson, 1901), pp. 38–41 and 391–411. On the early-twentieth century shift toward a less utilitarian discourse: F. Schmoll, 'Indication and Identification: On the History of Bird Protection in Germany, 1800–1918', in T. Lekan and T. Zeller (eds), *Germany's Nature. Cultural Landscapes and Environmental History* (New Brun-

swick: Rutgers University Press, 2005), pp. 161–82.

19. R. Boardman, *The International Politics of Bird Conservation: Biodiversity, Regionalism and Global Governance* (Cheltenham: Edward Elgar Publishing, 2006), pp. 39–40.

20. O. Kleinschmidt, 'Bericht über den Beschluss des V. Internationalen Zoologenkongresses', in M. Bedot (ed.) *Compte rendu des séances du sixième congrès internationale de zoologie tenu à Berne du 14 à 16 août 1904* (Geneva: W. Kündig & Fils, 1905), pp. 138–46, on p. 139.

21. As quoted in: D. E. Allen, *The Naturalist in Britain: A Social History* (Princeton, NJ: Princeton University Press, 1994), p. 178.

22. Some (particularly British) ornithologists showed themselves less averse to the society, and even became member in order to give it more of a scientific cachet. R. W. Doughty, *Feather Fashions and Bird Preservation: A Study in Nature Protection* (Berkeley, CA: University of California Press, 1975), pp. 52–3. See also: R. J. Moore-Colye, 'Feathered Women and Persecuted Birds: The Struggle against the Plumage Trade, c. 1860–1922', *Rural History*, 11:1 (2000), pp. 57–73.

23. Like some of the dissident ornithologists in Paris, he would point to the role of these animals in the equilibrium of nature's economy. Yet, they would end up anyway in a list of species of which 'it is desired to reduce the numbers'. For the entire text of the treaty: M. Cioc, *The Game of Conservation: International Treaties to Protect the World's Migratory Animals* (Athens, OH: Ohio University Press, 2009), pp. 154–161, on p. 161.

24. The ornithologist in question was Louis Ternier – a close friend of the conference's organizer. R. de Clermont, F. Cros-Mayrevieille and L. de Nussac (eds), *Le 1er congrès international pour la protection des paysages*. Compte rendu (Paris: Société pour la protection des paysages de France, 1910), pp. 133–4.

25. P. Sarasin, 'Über Weltnaturschutz', in R. Ritter von Stumer-Träunfels (ed.), *Verhandlungen des VIII. Internationalen Zoologen-Kongresses zu Graz* (Jena: Gustav Fischer Verlag 1910), pp. 240–53.

26. P. Sarasin, *Ueber die Aufgaben des Weltnaturschutzes. Denkschrift gelesen an der Delegiertenversammlung zur Weltnaturschutzkommission* (Basel: Verlag von Helbing & Lichtenbahn, 1914).

27. Sarasin, 'Über Weltnaturschutz', p. 240.

28. P. Sarasin, *Über die Ausrottung der Wal- und Robbenfauna sowie der Arktischen Tierwelt überhaupt* (Leipzig: August Pries, 1912).

29. Sarasin, 'Über Weltnaturschutz', pp. 245–54; Sarasin, *Ueber die Aufgaben*, pp. 8–55 and 61–2. On the international attempts to 'protect' *Naturvölker*, see: R. De Bont, '"Primitives" and Protected Areas: International Conservation and the "Naturalization" of Indigenous People, ca. 1910–1975', *Journal of the History of Ideas*, to be published in 2015.

30. On the history of the biocoenosis concept: K. Reise, 'Hundert Jahre Biocönose', *Naturwissenschaftliche Rundschau*, 33 (1980), pp. 328–35.

31. Sarasin, 'Über Weltnaturschutz', pp. 240 and 243–4.

32. See: Wöbse, *Weltnaturschutz*, pp. 50–2; Kupper, *Wildnis schaffen*, pp. 81–2.

33. Sarasin, *Ueber die Aufgaben*, pp. 11–13.

34. See in detail: Wöbse, *Weltnaturschutz*, pp. 54–8.

35. R. de Clermont et al. (eds), *Premier Congrès international pour la protection de la nature: faune et flore, sites et monuments naturels* (Paris: Société nationale de l'acclimatation de France, 1925).

36. A. Gruvel, G. Petit and C. Valois (eds), *Deuxième congrès international pour la protection de la nature* (Paris: Société d'éditions géographiques, maritimes et coloniales, 1932).

37. *La Protection de la Nature et l'Union Internationale des Sciences Biologiques* (Brussels:Office international pour la protection de la nature, 1929); 'International Conference for the Protection of the Fauna and Flora of Africa, 1933', Royal Belgian Institute of Natural Sciences (hereafter RBINS), typescript; 'Second International Conference for the Protection of the Fauna and Flora of Africa, 1938', RBINS, typescript.

38. E. Pelzers, *Geschiedenis van de Nederlandse commissie voor internationale natuurbescherming, de Stichting tot internationale natuurbescherming en het Office international pour la protection de la nature.* (Amsterdam: Nederlandse commissie voor internationale natuurbescherming, 1994); Barrow, *Nature's Ghosts*, pp. 142–3.

39. De Clermont, Cros-Mayrevieille and de Nussac (eds), *Le 1er Congrès*, pp. 5–9 and 154–5; De Clermont et al. (eds), *Premier Congrès*, pp. 6–29 and 381–6.

40. Six out of these were naturalists. Besides them the team was composed of game wardens, colonial administrators and a juridical adviser. 'International Conference, 1933', RBINS, typescript *Final Act*, p. 4.

41. This is particularly clear with regard to Belgium, a country that tried to profile itself in the interwar years with regard to scientific conservation in the colonies. Belgian scientists clearly had the ear (and access to the network) of the London ambassador and Baron Émile de Cartier de Marchienne, Prince Eugène de Ligne and the Crown Prince Leopold. This becomes clear, amongst other things, in Van Tienhoven's correspondence with his Belgian contacts; see: Amsterdam City Archives, Archief van de Nederlandse Commissie voor Internationale Natuurbescherming (hereafter ANCIN), 1283.

42. M. E. Keck and K. Sikkink, *Activists Beyond Borders: Advocacy Networks in International Politics* (Ithaca, NY: Cornell University Press, 1998).

43. Pieter Gerbrand van Tienhoven to Victor van Straelen, 24 December 1937, ANCIN, 1283–127.

44. Van Tienhoven to Van Straelen, 28 August 1935, ANCIN, 1283–127.

45. 'International conference, 1933', RBINS, typescript, 3rd session, pp. 12–13; 10th session, p. 14.

46. 'International conference, 1933', RBINS, typescript, 7th session, p. 2.

47. C. Bressou, 'La réserve de Camargue', in Gruvel, Petit and Valois (eds), *Deuxième congrès*, pp. 463–70.

48. M. Aullo, J. M. Priego and O. de Buen, 'La protection de la nature en Espagne', in Gruvel, Petit and Valois (eds), *Deuxième congrès*, pp. 69–77.

49. With regard to the shifting prestige: K. R. Benson, 'From Museum Research to Laboratory Research. The Transformation of Natural History into Academic Biology', in R. Rainer, K. R. Benson and J. Maienschain (eds), *The American Development of Biology* (New Brunswick: Rutgers University Press, 1991), pp. 49–83; L. Nyhart, 'Natural History and the "New" Biology', in N. Jardine, J. A. Secord, and E. C. Spary (eds), *Cultures of Natural History* (Cambridge: Cambridge University Press, 1996), pp. 426–46.

50. Good examples include Abel Gruvel (Paris), Percy Lowe (London), Victor van Straelen (Brussels) and Jan Sztolcman (Warsaw).

51. De Clermont et al. (eds), *Premier congrès*, pp. 352–9; A. Gruvel, *Titres et travaux scientifiques* (Paris: Société d'éditions géographiques, maritimes et coloniales,1930), p. 14.

52. In the early twentieth century this turn to conservation was limited to a few initiatives. Only in the 1970s would the rise of ecology lead to a more general trend. See E. Baratay and E. Hardouin-Fugier, *Zoo: A History of Zoological Gardens in the West* (Clerkenwell: Reaktion Books, 2004), pp. 145 and 235–6; W. M. Adams, *Against Extinction: The Story of Conservation* (London and New York, NY: Earthscan, 2004), pp. 137–40; J.

C. Donahue and E. K. Trump, *American Zoos During the Depression: A New Deal for Animals* (Jefferson, NC: McFarland, 2010), pp. 59–65.

53. Barrow, *Nature's Ghosts*, pp. 106–34.
54. Priemel to W. Reid Blair, 7 December 1930 and 3 Decemeber 1931, ANCIN, 1283–192.
55. Blair to Priemel, 19 December 1931, ANCIN, 1283–192.
56. L. Heck, 'Deutsche Fachschaft der Wisent-Züchter und-Halter', reprint from *Deutsche Jagd*, 22 (1934), ANCIN, 1283–192.
57. Heck to Van Tienhoven, 30 October 1939, ANCIN, 1283–192.
58. See: S. Schama, *Landscape and Memory* (Waukegan: Vintage, 1996), pp. 67–72; D. Ackerman, *The Zookeeper's Wife: A War Story* (New York, NY: W.W. Norton, 2007).
59. See extensively in: A. Runte, *National Parks: The American Experience*, 4th edn (Plymouth: Taylor Trade Publishing, 2010).
60. J.-M. Derscheid, 'La protection de la nature au Congo belge', in *La Protection de la Nature*, pp. 39–47; 'International conference, 1933', RBINS, typescript, 3rd session, pp. 5–6.
61. The Belgian ambassador in London, Emile Cartier de Marchienne, wrote as early as 1929 to his friend Van Tienhoven that he was pressuring for an 'internationalization of the Kivu district'. Cartier de Marchienne to Van Tienhoven, 5 August 1929, ANCIN, 1283–47.
62. *La Protection de la Nature et l'Union Internationale*, p. 338.
63. T. Graim, 'La coopération internationale pour la protection de la nature', in Gruvel, Petit and Valois (eds), *Deuxième congrès*, pp. 330–40.
64. Van Tienhoven to Herbert Smith, 20 May 1925, ANCIN, 1283–74.
65. Van Tienhoven to Jeanne Derscheid, 16 September 1926, ANCIN, 1283–55.
66. Jeanne Derscheid to Van Tienhoven, 11 November 1926, ANCIN, 1283–55.
67. On this hunting etiquette in the British Empire: J. M. MacKenzie, *The Empire of Nature: Hunting, Conservation and British Imperialism* (London and Manchester: Manchester University Press, 1988).
68. Tienhoven to the direction of Koninklijke Nederlandsch-Indische Luchtvaart Maatschappij (KNILM), 21 February 1939, ANCIN, 1283–183; Van Tienhoven to Van Straelen, 8 February 1938, ANCIN, 1283–127.
69. Van Straelen to Van Tienhoven, 19 March 1937 and 22 June 1937, ANCIN, 1283–127.
70. L. Desnues, 'Réserves et zones de protection pour mammifères et oiseaux', in De Clermont et al. (eds), *Premier Congrès*, pp. 70–4.
71. Van Tienhoven to Georges van Havre, 11 March 1931, ANCIN, 1283–202.
72. 'International conference, 1933', RBINS, typescript, 10th session, pp. 18–22.
73. The quote is Tienhoven's, and refers to the American, Henry A. Snow, who – after having been denied access to Africa – tried his luck in the East Indies. Tienhoven to Simon de Graaff, 17 January 1924, ANCIN, 1283–184; Tienhoven to Jean-Marie Derscheid, 25 April 1925, ANCIN, 1283–55.
74. Good examples of the latter include Jean Delacour (in France), Jean-Marie Derscheid (in Belgium) and Frans Ernst Blaauw (in the Netherlands).
75. Van Tienhoven to Abel Gruvel, 10 February 1932, ANCIN, 1283–184.
76. On these developments: McCormick, *Reclaiming Paradise*; M. Holdgate, *The Green Web: A Union for World Conservation* (London: Earthscan, 1999); A. Schwarzenbach, *Saving the World's Wildlife: WWF – The First 50 Years* (Gland: WWF International, 2011); A.-K. Wöbse, '"The World After All Was One": The International Environmental Network of UNESCO and IUPN, 1945–1950', *Contemporary European History*, 20 (2011), pp. 331–48.

77. WWF International, *Indigenous Peoples and Conservation: WWF Statement of Princi-ples* (Gland: WWF, 2008).
78. Jeanne Derscheid to Van Tienhoven, 11 November 1926, ANCIN, 1283-55.

4 The Hour of the Experts?: Reflections on the Rise of Experts in Interbellum Europe

1. F. A. Hayek, *The Road to Serfdom: Text and Documents; The Definitive Edition* (Chi-cago, IL: University of Chicago Press, 2007), p. 45.
2. F. A. von Hayek, 'The Use of Knowledge in Society', *American Economic Review*, 35 (1945), pp. 521–30, on p. 522.
3. Hayek, *The Road to Serfdom*, p. 57.
4. Hayek explicitly refers to the alleged successes of NS-planning, such as the Autobahn, in order to deconstruct them. See generally on planning: D. van Laak, 'Planung. Ge-schichte und Gegenwart des Vorgriffs auf die Zukunft', *Geschichte und Gesellschaft*, 34 (2008), pp. 305–26.
5. On the role of technical experts in the Second World War: P. Kennedy, *The Engineers of Victory* (New York, NY: Harper Collins, 2012). Highlighting the considerable dif-ferences between Germany and the United States: K. K. Patel, *Soldiers of Labor: Labor Service in Nazi Germany and New Deal America, 1933–1945* (New York, NY: Cam-bridge University Press 2005); W. Schivelbusch, *Entfernte Verwandtschaft: Faschismus, Nationalsozialismus, New Deal 1933–1939* (Munich: Hanser, 2005).
6. N. Vossoughian, *Otto Neurath. The Language of the Global Polis* (Rotterdam: NAi Publ., 2008); O. Neurath, M. Eve and C. Burke, *From Hieroglyphics to Isotype: A Visual Autobiography* (London: Hyphen Press, 2010); F. Hartmann, *Bildersprache: Otto Neur-ath Visualisierungen* (Vienna: Wiener Universitätsverlag, 2006).
7. This is most obvious in the idea of a technocracy: D. van Laak, 'Technokratie im Eu-ropa des 20. Jahrhunderts – eine einflussreiche "Hintergrundideologie"', in L. Raphael (ed.), *Theorien und Experimente der Moderne. Europas Gesellschaften im 20. Jahrhun-dert* (Cologne: Böhlau 2012), pp. 101–28.
8. D. Pestre, 'Regimes of Knowledge Production in Society: Towards a More Political and Social Reading', *Minerva*, 41 (2003), pp. 245–62.
9. M. Szöllösi-Janze, 'Science and Social Space: The Transformations in the Institutions of Wissenschaft from the Wilhelmine Empire to the Weimar Republic', *Minerva*, 43 (2005), pp. 339–360.
10. A. Schlimm, 'Handeln im Angesicht der Krise: Zukunftswissen und Expertise deutscher Verkehrswissenschaftler in der ersten Hälfte des 20. Jahrhunderts', in H. Hartmann and J. Vogel (eds), *Zukunftswissen: Prognosen in Wirtschaft, Politik und Gesellschaft seit 1900* (Frankfurt: Campus, 2010), pp. 175–94.
11. L. Raphael, 'Die Verwissenschaftlichung des Sozialen als methodische und konzeptuelle Herausforderung für eine Sozialgeschichte des 20. Jahrhunderts', *Geschichte und Gesell-schaft*, 22 (1996), pp. 165–93; R. M. MacLeod, *Government and Expertise: Specialists, Administrators and Professionals, 1860–1919* (Cambridge: Cambridge University Press, 2003).
12. C. S. Maier, 'Consigning the Twentieth Century to History: Alternative Narratives for the Modern Era', *The American Historical Review*, 105 (2000), pp. 807–31; see also J. C. Scott, *Seeing Like a State: How Certain Schemes to Improve the Human Condition Have*

Failed (New Haven, CT: Yale University Press, 1998). Generally on this phenomenon, see C. Otter, *The Victorian Eye: A Political History of Light and Vision in Britain, 1800–1910* (Chicago, IL: University of Chicago Press, 2008).

13. A. Gouldner, *The Future of Intellectuals and the Rise of the New Class* (New York, NY: Seabury Press, 1979), p. 19.

14. E. Kurz-Milcke, 'The Authority of Representations', in E. Kurz-Milcke and G. Gigerenzer (eds), *Experts in Science and Society* (New York, NY: Kluwer Academic/Plenum Publishers, 2004), pp. 281–301.

15. C. MacLeod, *Heroes of Invention: Technology, Liberalism and British Identity, 1750–1914* (Cambridge: Cambridge University Press, 2007).

16. J. Schot and V. Lagendijk, 'Technocratic Internationalism in the Interwar Years: Building Europe on Motorways and Electricity Networks', *Journal of Modern European History*, 6 (2008), pp. 196–218.

17. M. Kohlrausch, 'Technologische Innovation und transnationale Netzwerke: Europa zwischen den Weltkriegen', *Journal of Modern European History*, 6 (2008), pp. 181–95; H. Rakel, 'Scientists as Expert Advisors: Science Cultures Versus National Cultures', in E. Kurz-Milcke and G. Gigerenzer (eds), *Experts in Science and Society* (New York, NY: Kluwer Academic/Plenum Publishers, 2004), pp. 3–25.

18. A. F. Frank, *Oil Empire: Visions of Prosperity in Austrian Galicia* (Cambridge, MA: Harvard University Press, 2005).

19. M. Kohlrausch, K. Steffen and S. Wiederkehr (eds), *Expert Cultures in Central Eastern Europe: The Internationalization of Knowledge and the Transformation of Nation States Since World War I* (Osnabrück: Fibre, 2010).

20. Generally: A. R. Seipp, *The Ordeal of Peace: Demobilization and the Urban Experience in Britain and Germany, 1917–1921* (Farnham: Ashgate, 2009).

21. M. Bucur, *Eugenics and Modernization in Interwar Romania* (Pittsburgh, PA: University of Pittsburgh Press, 2002); P. Weindling, 'Public Health and Political Stabilization: The Rockefeller Foundation in Central and Eastern Europe Between the Two World Wars', *Minerva*, 31 (1993), pp. 253–67.

22. K. Steffen, 'Wissenschaftler in Bewegung: Der Materialforscher Jan Czochralski zwischen den Weltkriegen', *Journal of Modern European History*, 6 (2008), pp. 237–261.

23. Quoted in G. Péteri, 'Engineer Utopia. On the Position of Technostructure in Hungary's War Communism, 1919', *International Studies of Management & Organization*, 19 (1989), pp. 82–101, on p. 87.

24. Péteri, 'Engineer Utopia', p. 91.

25. T. Veblen, *The Engineers and the Price System* (1921) (Kitchener: Batoche Books, 2001).

26. Quoted in Veblen, *The Engineers*, pp. 93–4.

27. J. Piłatowicz, *Kadra Inżynierska w II Rzeczypospolitej* (Siedlce: Wydawnictwo Wyższej Szkoły Rolniczo-Pedagogicznej, 1994), pp. 123–8.

28. Quoted in: E. van Meer, '"The Nation is Technological": Technical Expertise and National Competition in the Bohemian Lands, 1880–1914', in Kohlrausch, Steffen and Wiederkehr (eds), *Expert Cultures*, pp. 85–104, on p. 102.

29. M. Efmertová, 'Les Professeurs Électrotechnicies Tchèques dans le Monde: Formation et Impact des Travaux Scientifiques dans les Années 1918–1938', in A. de Cardoso Matos (ed.), *The Quest for a Professional Identity: Engineers between Training and Action* (Lisbon: Colibri, 2009), pp. 513–23.

30. S. Rohdewald, 'Mimicry in a Multiple Postcolonial Setting: Networks of Technocracy and Scientific Management in Piłsudskis Poland', in Kohlrausch, Steffen and

Wiederkehr (eds), *Expert Cultures*, pp. 63–84; M. M. Drozdowski, *Eugeniusz Kwiatkowski* (Rzeszów: Wysza Szkoła Informatyki i Zarządzania, 2005).

31. J. Żarnowski, 'Learned Professions in Poland 1918–1939', in J. Żarnowski (ed.), *State, Society and Intelligentsia: Modern Poland and its regional context* (Aldershot:Ashgate Variorum, 2003), pp. 407–26, on p. 413.

32. I. Loose, 'How to Run a State: The Question of Knowhow in Public Administration in the First Years after Poland's Rebirth in 1918', in Kohlrausch, Steffen and Wiederkehr (eds), *Expert Cultures*, pp. 145–59.

33. E. Forsthoff, *Die Verwaltung als Leistungsträger* (Stuttgart: Kohlhammer, 1938); F. Meinel, *Der Jurist in der industriellen Gesellschaft. Ernst Forsthoff und seine Zeit* (Berlin: Akademie Verlag, 2011).

34. See T. Etzemüller, 'Social Engineering als Verhaltenslehre des kühlen Kopfes', in T. Etzemüller (ed.), *Die Ordnung der Moderne: Social engineering im 20. Jahrhundert* (Bielefeld: Transcript, 2009).

35. C. Hirschi, *The Origins of Nationalism. An Alternative History from Ancient Rome to Early Modern Germany* (Cambridge: Cambridge University Press, 2012), p. 31.

36. T. Todorov, *The Limits of Art: Two Essays* (London: Seagull Books, 2010), p. 42.

37. Quote from the 1930s, reprinted in: F. Schumacher (ed.), *Lesebuch für Baumeister. Äußerungen über Architektur und Städtebau: Eine Sammlung klassischer Texte* (1948) (Braunschweig: Vieweg, 1977).

38. A. Behne, 'Dammerstock', *Die Form. Zeitschrift für gestaltende Arbeit*, 5 (1930), pp. 163–6, on p. 164.

39. Letter from Le Corbusier to S. Giedion, 29 August 1933, reprinted in: T. Hilpert, *Le Corbusiers 'Charta von Athen'. Texte und Dokumente* (Braunschweig: Vieweg, 1984), p. 171.

40. S. Dierig, J. Lachmund and A. W. Mendelsohn, 'Introduction: Toward an Urban History of Science', *Osiris: Science and the Rise of Modern Cities*, 18 (2003), pp. 1–22.

41. M. Straalen, 'Empirische Stadtanalysen', *Daedalus*, 69/70 (1998), pp. 60–7.

42. E. Blau, *The Architecture of Red Vienna, 1919–1934* (Cambridge, MA: MIT Press, 1999).

43. S. Giedion, *Space, Time & Architecture: The Growth of a New Tradition* (Cambridge, MA: Harvard University Press, 1941); S. Giedion, *Mechanization Takes Command. A Contribution to Anonymous History* (Oxford: Oxford University Press, 1948).

44. M. Callon, 'Some Elements of a Sociology of Translation: Domestication of the Scallops and the Fishermen of St Brieuc Bay', in J. Law, *Power, Action and Belief: A New Sociology of Knowledge* (London: Routledge, 1986), pp. 196–223.

45. A. Saint, *The Image of the Architect* (New Haven, CT: Yale University Press, 1983).

46. http://moholy-nagy.org/ [accessed 19 January 2015].

47. S. Giedion, *Sigfried Giedion und die Fotografie Bildinszenierungen der Moderne* (Zürich: gta-Verlag, 2010); J. Sbriglio, *Le Corbusier & Lucien Herve: A Dialogue Between Architect and Photographer* (Los Angeles, CA: J. Paul Getty Museum, 2011); S. B. Uzelac, 'Visual Arts in the Avant-Gardes Between the Two Wars', in D. Djurić and M. Šuvaković (eds), *Impossible Histories: Historical Avant-Gardes, Neo-Avant-Gardes, and Post-Avant-Gardes in Yugoslavia, 1918–1991* (Cambridge, MA: MIT Press, 2003), pp.122–69.

48. B. Rieger, *Technology and the Culture of Modernity in Britain and Germany. 1890–1945* (Cambridge, Cambridge University Press, 2005); A. C. T. Geppert, 'Space Personae: Cosmopolitan Networks of Peripheral Knowledge 1927–1957', *Journal of Modern European History*, 6 (2008), pp. 262–8.

49. R. Jessen and J. Vogel, 'Die Naturwissenschaften und die Nation: Perspektiven einer

Wechselbeziehung in der europäischen Geschichte', in R. Jessen and J. Vogel (eds), *Wissenschaft und Nation in der europäischen Geschichte* (Frankfurt: Campus-Verlag, 2002), pp. 7–37; and in general: MacLeod, *Heroes of Invention*.

50. D. Augustine, *Red Prometheus. Engineering and Dictatorship in East Germany, 1945–1990* (Cambridge, MA: MIT Press, 2007).

51. S. Brukalski, 'Popularność i publiczność architektury', in T. Barucki (ed.), *Fragmenty Stuletniej Historii. 1899–1999: Relacje, Wspomnenia, Refleksje w Stulecie Organizacji Warszawskich Architektów* (Warsaw: Drukoba, 2000), pp. 81–3.

52. H. Bodenschatz and C. Post (eds), *Städtebau im Schatten Stalins: Die internationale Suche nach der sozialistischen Stadt in der Sowjetunion 1929–35* (Berlin: Braun, 2003), pp. 30–43.

5 The Psychiatrist as the Leader of the Nation: Psycho-Political Expertise after the German Revolution, 1918–19

1. V. Roelcke, *Krankheit und Kulturkritik: Psychiatrische Gesellschaftsdeutungen im bürgerlichen Zeitalter* (Frankfurt: Campus, 1999), pp. 204–5.

2. See also D. Freis, 'Die "Psychopathen" und die "Volksseele": Psychiatrische Diagnosen des Politischen und die Novemberrevolution 1918/1919', in H.-W. Schmuhl and V. Roelcke (eds), *'Heroische Therapien': Die deutsche Psychiatrie im internationalen Vergleich 1918–1945* (Göttingen: Wallstein, 2013), pp. 48–68.

3. M. Föllmer, 'Der "kranke Volkskörper": Industrielle, hohe Beamte und der Diskurs der nationalen Regeneration in der Weimarer Republik', *Geschichte und Gesellschaft*, 27 (2001), pp. 41–67.

4. R. Graf, 'Optimismus und Pessimismus in der Krise: Der politisch-kulturelle Diskurs in der Weimarer Republik', in W. Hardtwig (ed.), *Ordnungen in der Krise: Zur politischen Kulturgeschichte Deutschlands 1900–1933* (Munich: Oldenbourg, 2007), pp. 115–40.

5. V. Roelcke, 'Die Entwicklung der Psychiatrie zwischen 1880 und 1932: Theoriebildung, Institutionen, Interaktionen mit zeitgenössischer Wissenschafts- und Sozialpolitik', in R. vom Bruch and B. Kaderas (eds), *Wissenschaften und Wissenschaftspolitik: Bestandsaufnahmen zu Formationen, Brüchen und Kontinuitäten im Deutschland des 20. Jahrhunderts* (Stuttgart: F. Steiner, 2002), pp. 109–24.

6. P. Lerner, *Hysterical Men: War, Psychiatry, and the Politics of Trauma in Germany, 1890–1930* (Ithaca, NY: Cornell University Press, 2003), pp. 209–14.

7. 'Psychiatrisches Ausbreitungsbedürfnis', *Die Irrenrechts-Reform*, 64 (1919), pp. 197–9.

8. M. Föllmer, R. Graf and P. Leo, 'Die Kultur der Krise in der Weimarer Republik', in M. Föllmer and R. Graf (eds), *Die "Krise" der Weimarer Republik: Zur Kritik eines Deutungsmusters* (Frankfurt and New York, NY: Campus, 2005), pp. 9–41; W. Hardtwig, 'Einleitung: Politische Kulturgeschichte der Zwischenkriegszeit', in W. Hardtwig (ed.), *Politische Kulturgeschichte der Zwischenkriegszeit 1918–1939* (Göttingen: Vandenhoeck & Ruprecht, 2005), pp. 7–22, on p. 8.

9. L. Raphael, 'Die Verwissenschaftlichung des Sozialen als methodische und konzeptionelle Herausforderung für eine Sozialgeschichte des 20. Jahrhunderts', *Geschichte und Gesellschaft*, 22:2 (1996), pp. 165–93.

10. E. Kahn, 'Psychopathie und Revolution', *Münchener Medizinische Wochenschrift*, 66:43 (1919), pp. 968–9, on p. 986.

11. H. Hippius et al., *Die Psychiatrische Klinik der Universität München 1904–2004* (Hei-

delberg: Springer, 2005), p. 90.

12. E. Kahn, 'Psychopathen als revolutionäre Führer', *Zeitschrift für die gesamte Neurologie und Psychiatrie*, 52:1 (1919), pp. 90–106, on p. 90.

13. Kahn, 'Psychopathen', p. 91.

14. P. Riedesser and A. Verderber, *'Maschinengewehre hinter der Front': Zur Geschichte der deutschen Militärpsychiatrie* (Frankfurt: Fischer, 1996), pp. 82–5.

15. Kahn, 'Psychopathie', p. 968.

16. R. Wetzell, *Inventing the Criminal: A History of German Criminology, 1880–1945* (Chapel Hill, NC and London: University of North Carolina Press, 2000), pp. 145–9.

17. Kahn, 'Psychopathen' p. 92; E. Kraepelin, *Psychiatrie: Ein Lehrbuch für Studierende und Ärzte*, 7th edn, 2 vols (Leipzig: J.A. Barth, 1903), vol. 2, pp. 815–41.

18. Kahn, 'Psychopathie', p. 968.

19. See, for example, K. Hildebrandt, 'Forensische Begutachtung eines Spartakisten', *Allgemeine Zeitschrift für die Psychiatrie und psychisch-gerichtliche Medizin*, 76 (1918), pp. 489–518; H. Stelzner, 'Psychopathologisches in der Revolution', *Zeitschrift für die gesamte Neurologie und Psychiatrie*, 49 (1919), pp. 393–408; H. Freimark, *Die Revolution als psychische Massenerscheinung: Historisch-Psychologische Studie* (Munich and Wiesbaden: J.F. Bergmann, 1920), pp. 64–97.

20. Wetzell, *Inventing the Criminal*, p. 145; U. Germann, *Psychiatrie und Strafjustiz: Entstehung, Praxis und Ausdifferenzierung der forensischen Psychiatrie in der deutschsprachigen Schweiz 1850–1950* (Zurich: Chronos, 2004), p. 470.

21. Lerner, *Hysterical Men*, p. 210.

22. Lerner, *Hysterical Men*, p. 214.

23. Lerner, *Hysterical Men*, p. 214.

24. D. Kaufmann, 'Science as Cultural Practice: Psychiatry in the First World War and Weimar Germany', *Journal of Contemporary History*, 34:1 (1999), pp. 125–44, on p. 142.

25. H. Brennecke, 'Zur Frage der Psychopathologie der Revolution und der Revolutionäre', *Zeitschrift für Kinderforschung*, 26:5 (1921), pp. 225–31, on p. 228.

26. Brennecke, 'Zur Frage', pp. 229–30; C. Müller, *Verbrechensbekämpfung im Anstaltsstaat: Psychiatrie, Kriminologie und Strafrechtsreform in Deutschland 1871–1933* (Göttingen: Vandenhoeck & Ruprecht, 2004), pp. 126–49.

27. Kahn, 'Psychopathie', p. 969.

28. Kahn, 'Psychopathen', p.105.

29. Kahn, 'Psychopathie', p. 969; on German psychiatry as a 'belated science' see: D. Blasius, *Einfache Seelenstörung: Geschichte der deutschen Psychiatrie 1800–1945* (Frankfurt: Fischer, 1994), p. 117.

30. R. Sommer, *Ärztlicher Notruf zum Ende des Jahres 1918* (Gießen: Published by the author, 1918).

31. K. Bonhoeffer, 'Inwieweit sind politische und soziale und kulturelle Zustände einer psychopathologischen Betrachtung zugänglich?', *Klinische Wochenschrift*, 2:13 (1923), pp. 598–601.

32. S. Freud, 'Zeitgemäßes über Krieg und Tod' (1915), in A. Freud et al. (eds), *Sigmund Freud: Gesammelte Werke*, (Frankfurt: Fischer, 1980), vol. 10, pp. 324–355, on p. 324.

33. 'Massenwahnsinn?' *Die Irrenrechts-Reform*, 63 (1919), p. 181–4.

34. T. Mergel, 'Führer, Volksgemeinschaft und Maschine: Politische Erwartungsstrukturen in der Weimarer Republik und dem Nationalsozialismus 1918–1936', in W. Hardtwig (ed.), *Politische Kulturgeschichte der Zwischenkriegszeit 1918–1939* (Göttingen: Vandenhoeck & Ruprecht, 2005), pp. 91–128.

35. Kahn, 'Psychopathen', p. 101.
36. Kahn, 'Psychopathen', pp. 101–2.
37. Kahn, 'Psychopathen', p. 103.
38. See also W. Hardtwig, 'Der Bismarck-Mythos: Gestalt und Funktionen zwischen politischer Öffentlichkeit und Wissenschaft', in W. Hardtwig (ed.), *Politische Kulturgeschichte der Zwischenkriegszeit 1918–1939* (Göttingen: Vandenhoek & Ruprecht, 2005), pp. 61–90.
39. E. J. Engstrom, 'Emil Kraepelin: Psychiatry and Public Affairs in Wilhelmine Germany', *History of Psychiatry*, 2:6 (1991), pp. 111–32, on pp. 131–2.
40. R. Gaupp, 'Der nervöse Zusammenbruch und die Revolution', *Blätter für Volksgesundheitspflege*, 19:5/6 (1919), pp. 43–6, on p. 43.
41. Gaupp, 'Zusammenbruch', p. 45.
42. Gaupp, 'Zusammenbruch', p. 46.
43. R. Gaupp, *Die zukünftige Stellung des Arztes im Volke: Ansprache an die Studierenden der Medizin der Universität Tübingen (23. X. 1919)* (Tübingen: H. Laupp, 1919), p. 4.
44. Gaupp, *Stellung*, p. 20.
45. Gaupp, *Stellung*, pp. 14–8.
46. Mergel, 'Führer', p. 105.
47. Gaupp, *Stellung*, p. 20.
48. E. Kraepelin, 'Psychiatrische Randbemerkungen zur Zeitgeschichte', *Süddeutsche Monatshefte*, 16 (1919), pp. 171–183, on p. 175; Engstrom, 'Kraepelin', pp. 128–132; on Kraepelin's scientific and political thought see also: M. Weber, W. Burgmair and E. J. Engstrom, 'Emil Kraepelin: Zwischen klinischen Krankheitsbildern und "psychischer Volkshygiene"', *Deutsches Ärzteblatt*, 103:41 (2006), pp. 2685–90; V. Roelcke, 'Biologizing Social Facts: An Early 20th Century Debate on Kraepelin's Concept of Culture, Neurasthenia, and Degeneration', *Culture, Medicine, and Psychiatry*, 21:4 (1997), pp. 383–403.
49. Kraepelin, 'Randbemerkungen', translated and cited in Engstrom, 'Kraepelin', p. 129, original emphasis.
50. Kraepelin, 'Randbemerkungen', p. 176.
51. Engstrom, 'Kraepelin', pp. 129–30.
52. Engstrom, 'Kraepelin', p. 130.
53. Kraepelin, 'Randbemerkungen', p. 176.
54. Kraepelin, 'Randbemerkungen', p. 180.
55. Kraepelin, 'Randbemerkungen', p. 181.
56. Engstrom, 'Kraepelin', p. 131.
57. Kraepelin, 'Randbemerkungen', p. 182; S. Kühl, *Die Internationale der Rassisten: Aufstieg und Niedergang der internationalen Bewegung für Eugenik und Rassenhygiene im 20. Jahrhundert* (Frankfurt and New York, NY: Campus, 1997), pp. 41–58.
58. Kraepelin, 'Randbemerkungen', pp. 182–3.
59. Kraepelin, 'Randbemerkungen', pp. 182–3; H. L. Siemen, 'Reform und Radikalisierung: Veränderungen der Psychiatrie in der Weltwirtschaftskrise', in N. Frei (ed.), *Medizin und Gesundheitspolitik in der NS-Zeit* (Munich: Oldenbourg, 1991), pp. 191–200.
60. E. Stransky, 'Angewandte Psychiatrie: Motive und Elemente zu einem Programmentwurf', *Allgemeine Zeitschrift für Psychiatrie und psychisch-gerichtliche Medizin*, 74:1–3 (1918), pp. 22–53, on pp. 22–3.
61. Stransky, 'Angewandte Psychiatrie', p. 30.
62. Stransky, 'Angewandte Psychiatrie', p. 31.

63. Stransky, 'Angewandte Psychiatrie', p. 35.
64. Stransky, 'Angewandte Psychiatrie', p. 37.
65. Stransky, 'Angewandte Psychiatrie', p. 44.
66. E. Stransky, 'Der seelische Wiederaufbau des deutschen Volkes und die Aufgaben der Psychiatrie', *Zeitschrift für die gesamte Neurologie und Psychiatrie*, 60 (1920), pp. 271–80, on p. 277.
67. E. Stransky, 'Wiederaufbau'.
68. A. Kronfeld, 'Die Bedenklichkeit der "angewandten" Psychiatrie', *Zeitschrift für die gesamte Neurologie und Psychiatrie*, 65 (1921), pp. 364–7, on p. 366.
69. I.-W. Kittel, 'Arthur Kronfeld zur Erinnerung – Schicksal und Werk eines jüdischen Psychiaters in drei deutschen Reichen', in I.-W. Kittel (ed.), *Arthur Kronfeld (1886–1941): Ein Pionier der Psychologie, Sexualwissenschaft und Psychotherapie* (Konstanz: Bibliothek der Universität Konstanz, 1988), pp. 7–14, on p. 8.
70. K. Birnbaum, *Grundzüge der Kulturpsychopathologie* (Munich: J. Bergmann, 1924); A. Kronfeld, *Das seelisch Abnorme und die Gemeinschaft* (Stuttgart: J. Püttmann, 1923).
71. S. Freud, *Massenpsychologie und Ich-Analyse. Die Zukunft einer Illusion* (Frankfurt: Fischer, 1993).
72. Föllmer, 'Volkskörper'.
73. Roelcke, *Krankheit*; see also J. Radkau, *Das Zeitalter der Nervosität: Deutschland zwischen Bismarck und Hitler* (Munich: Hanser, 1998).
74. H. Oppenheim 'Seelenstörung und Volksbewegung', *Berliner Tageblatt*, 48:171 (16 April 1919), pp. 1–2.
75. Roelcke, 'Entwicklung', p. 119.
76. E. Kraepelin, 'Ein Forschungsinstitut für Psychiatrie', *Zeitschrift für die gesamte Neurologie und Psychiatrie*, 32 (1916), pp. 1–38, on p. 2.
77. Kraepelin, 'Eine Forschungsinstitut', p. 2.
78. M. M. Weber, 'Psychiatric Research and Science Policy in Germany: The History of the Deutsche Forschungsanstalt Für Psychiatrie (German Institute for Psychiatric Research) in Munich from 1917 to 1945', *History of Psychiatry*, 11 (2000), pp. 235–58.
79. Weber, 'Psychiatric Research', p. 247.
80. M. M. Weber, 'Rüdin, Ernst', *Neue Deutsche Biographie* (2005), pp. 215–6.
81. D. J. K. Peukert, *Die Weimarer Republik: Krisenjahre der klassischen Moderne* (Frankfurt: Suhrkamp, 1987), pp. 132–40.
82. M. G. Ash, 'Wissenschaft und Politik als Ressourcen für einander', in R. vom Bruch and B. Kaderas (eds), *Wissenschaften und Wissenschaftspolitik: Bestandsaufnahmen zu Formationen, Brüchen und Kontinuitäten im Deutschland des 20. Jahrhunderts* (Stuttgart: Steiner, 2002), pp. 32–51.
83. S. Neuner, *Politik und Psychiatrie: Die staatliche Versorgung psychisch Kriegsbeschädigter in Deutschland 1920–1939* (Göttingen: Vandenhoeck & Ruprecht, 2011), pp. 91–110; see also Lerner, 'Hysterical Men', pp. 223–48.
84. Wetzell, *Inventing the Criminal*, p. 125.
85. R. F. Wetzell, 'Die Rolle medizinischer Experten in Strafjustiz und Strafrechtsreformbewegung: Eine Medikalisierung des Strafrechts?' in A. Kästner and S. Kesper-Biermann (eds), *Experten und Expertenwissen in der Strafjustiz von der frühen Neuzeit bis zur Moderne* (Leipzig: Meine, 2008), pp. 57–72, on pp. 66–7.
86. A. Stegerwald, 'Minister für Volkswohlfahrt an sämtliche Ober– und Regierungspräsidenten, 2 September 1920, Berlin', in F.W. Kersting and H.-W. Schmuhl (eds), *Quellen zur Geschichte der Anstaltspsychiatrie in Westfalen* (Paderborn: Schöningh, 2004), pp. 174–5.

87. M. Thomson, 'Mental Hygiene as an International Movement', in P. Weindling (ed.), *International Health Organizations and Movements, 1918–1939* (Cambridge: Cambridge University Press, 1995), pp. 283–304.

88. V. Roelcke, 'Prävention in Hygiene und Psychiatrie zu Beginn des 20. Jahrhunderts: Krankheit, Gesellschaft, Vererbung und Eugenik bei Robert Sommer und Emil Gotschlich', in U. Enke (ed.), *Die Medizinische Fakultät der Universität Gießen: Institutionen, Akteure und Ereignisse von der Gründung 1607 bis ins 20. Jahrhundert* (Stuttgart: F. Steiner, 2007), pp. 395–416, on p. 395.

89. P. Weindling, *Health, Race and German Politics Between National Unification and Nazism, 1870–1945* (Cambridge: Cambridge University Press, 1989), p. 307.

90. Siemen, 'Radikalisierung', pp. 196–9.

91. Roelcke, 'Entwicklung', pp. 121–4.

6 Contested Modernity: A. G. Doiarenko and the Trajectories of Agricultural Expertise in Late Imperial and Soviet Russia

1. This chapter is based on the results of my PhD thesis, published as *Als das Dorf noch Zukunft war. Agrarismus und Expertise zwischen Zarenreich und Sowjetunion* (Cologne: Böhlau, 2014).

2. Letter of Congratulation to A. G. Doiarenko, 1926, Russian State Economic Archive (hereafter RGAĖ), f. 9474, op. 1, d. 231, l. 1.

3. Letter of Congratulation, RGAĖ, f. 9474, op. 1, d. 231, l. 1.

4. S. K. Chaianov, 'Aleksei Grigor'evich Doiarenko (K 25-letnemu iubileiu)', *Sel'skokhoziaistvennaia Zhizn'*, 6:52 (1926), pp. 14–5, on p. 14.

5. A. G. Doiarenko, *Iz agronomicheskogo proshlogo* (Moscow: Gosudarstvennoe izdatel'stvo sel'skokhoziaistvennoi literatury, 1958), pp. 62–7.

6. A. G. Doiarenko, *Krest'ianskie besedy. Prakticheskoe polevodstvo na opytnom pole* (Moscow: Izdanie M. i S. Sabashnikovykh, 1925).

7. On A. G. Doiarenko's biography, see A. A. Kurenyshev, *On slyshal muzyku polei … Zhizn' i deiatel'nost' Alekseia Grigorevicha Doiarenko, uchenogo, pedagoga, obshchestven-nogo deiatelia, muzykanta. 1874 – 1958 gg.* (Moscow: AIRO XXI, 2011).

8. M. Dolbilov, 'The Emancipation Reform of 1861 in Russia and the Nationalism of the Imperial Bureaucracy', in T. Hayasti (ed.), *The Construction and Deconstruction of National Histories in Slavic Eurasia* (Sapporo: SRC, 2003), pp. 205–35.

9. C. A. Frierson, *Peasant Icons. Representations of Rural People in Late Nineteenth-Century Russia* (Oxford: Oxford University Press, 1993); W. S. Vucinich (ed.), *The Peasant in Nineteenth-Century Russia* (Stanford, CA: Stanford University Press, 1968); M. Hughes, 'Misunderstanding the Russian Peasantry. Anti-Capitalist Revolution or Third Rome? Interactions between Agrarianism, Slavophilism and the Russian *narodniki*', in H. Schultz and A. Harre (eds), *Bauerngesellschaften auf dem Weg in die Moderne. Agrarismus in Ostmitteleuropa 1880 – 1960* (Wiesbaden: Harrassowitz, 2010), pp. 55–67.

10. R. E. Johnson, 'Liberal Professionals and Professional Liberals. The Zemstvo Statisticians and their Work', in T. Emmons and W. S. Vucinich (eds), *The Zemstvo in Russia. An Experiment in Local Self-Government* (Cambridge: Cambridge University Press, 1982), pp. 343–63.

11. D. W. Darrow, 'From Commune to Household: Statistics and the Social Construction of Chaianov's Theory of Peasant Economy', *Comparative Studies in Society and History*,

43:4 (2001), pp. 788–818.

12. M. G. Pavlov cited in O. Iu. Elina, *Ot tsarskikh sadov do sovetskikh polei. Istoriia sel'skokhoziaistvennykh opytnykh uchrezhdenii XVIII – 20-e gody XX veka* (Moscow: Institut istorii, estestvoznaniia i tekhniki RAN, 2008.), vol. 1, p. 16.

13. A. F. Fortunatov, *Sel'skoe khoziaistvo i agronomiia* (Moscow: Tipo-lit I. N. Kushnerev i K°, 1903), p. 11.

14. E. Kingston-Mann, 'Statistics, Social Science, and Social Justice. The Zemstvo Statisticians of Pre-Revolutionary Russia', in S. P. McCaffray and M. Melancon (eds), *Russia in the European Context 1789–1914. A Member of the Family* (New York, NY: Palgrave Macmillian, 2005), pp. 113–39.

15. A. I. Chuprov, 'Agrarnaia reforma i ee veroiatnoe vliianie na sel'skokhoziaistvennoe proizvodstvo' (1906), in A. I. Chuprov (ed.), *Rossiia vchera i zavtra. Stat'i. Rechi. Vospominaniia* (Moscow: Russkii mir, 2009), pp. 238–56; V. A. Kosinskii, *K agrarnomu voprosu* (Odessa: 'Ėkonomicheskaia' tip., 1906); A. A. Manuilov, *Pozemel'nyi vopros v Rossii. Malozemel'e, dopol'nitel'nyi nadel i arenda* (Moscow: 1905).

16. This is a typical strategy of expert elites, see: R. Hitzler, 'Wissen und Wesen des Experten. Ein Annäherungsversuch– zur Einleitung', in R. Hitzler, A. Honer and C. Maeder (eds), *Expertenwissen. Die institutionalisierte Kompetenz zur Konstruktion von Wirklichkeit* (Opladen: Westdeutscher Verlag, 1994), pp. 13–30, on p. 17.

17. A. F. Fortunatov, 'Kto on?', *Zemskii Agronom*, 1 (1913), pp. 13–18, on p. 14.

18. A. V. Chaianov, *Chto takoe agrarnyi vopros?* (Moskva: Universal'naia biblioteka), p. 25. This publication is considered the manifesto of the League for Agrarian Reforms.

19. A. G. Doiarenko, 'Pod-em proizvoditel'nosti sel'skogo khoziaistva i agrarnaia reforma', *Vestnik sel'skogo khoziaistva*, 16 (1917), pp. 3–4, on p. 4.

20. On the topos of peasant backwardness in late imperial Russia, see Y. Kotsonis, *Making Peasants Backward. Agricultural Cooperatives and the Agrarian Question in Russia, 1861–1914* (New York, NY: St. Martin's Press, 1999).

21. G. L. Yaney, *The Urge to Mobilize. Agrarian Reform in Russia, 1861–1930* (Urbana, IL: University of Illinois Press, 1982); G. L. Yaney, 'Some Aspects of the Imperial Russian Government on the Eve of the First World War', *The Slavonic and East European Review*, 43:100 (1964), pp. 68–90.

22. D. A. Macey, 'Reflections on Peasant Adaptation in Rural Russia at the Beginning of the Twentieth Century. The Stolypin Agrarian Reforms', in S. K. Wegren (ed.), *Rural Adaptation in Russia* (London: Routledge, 2005), pp. 38–64.

23. I. V. Gerasimov, *Modernism and Public Reform in Late Imperial Russia. Rural Professionals and Self-Organization 1905–1930* (Houndmills: Palgrave Macmillan, 2009).

24. D. L. Hoffmann, *Cultivating the Masses. Modern State Practices and Soviet Socialism, 1914 – 1939* (Ithaca, NY: Cornell University Press, 2011); D. Beer, *Renovating Russia. The Human Sciences and the Fate of Liberal Modernity, 1880–1930* (Ithaca NY: Cornell University Press, 2008).

25. N. Stehr and R. Grundmann, *Experts. The Knowledge and Power of Expertise* (London and New York, NY: Routledge, 2011), p. x.

26. Gerasimov, *Modernism and Public Reform*, chapter 6.

27. A. G. Doiarenko, *K piatidesiatiletiiu Petrovskoi Akademii. 1865g. – 21. noiabria 1915g.* (Moscow: Kn-vo studentov M. S.-Ch. I., 1916), p. 8.

28. Gerasimov, *Modernism and Public Reform*, pp. 33–4.

29. Such public educational initiatives were a form of 'civil society in practice'. See D. Wartenweiler, *Civil Society and Academic Debate in Russia. 1905–1914* (Oxford: Clar-

endon Press, 1999), chapter 5.

30. *Otchet Golitsynskikh zhenskikh sel'skochoziaistvennykh kursov za ... god* (Moscow: Tipo-litografiia V. Richter, 1911–16).

31. Among those teaching were the agronomists A. F. Fortunatov, A. G. Doiarenko and A. P. Levitskii; the economists V. I. Anisimov, B. D. Bruckus, M. I. Tugan-Baranovskii, N. P. Makarov, S. N. Prokopovich, A. N. Minin, Z. S. Katsenelenbaum and L. B. Kafengauz; and the cooperative activists V. V. Khizhniakov and V. A. Kil'chevskii.

32. *Otchet Moskovskogo Gorodskogo Narodnogo Universiteta im. A. L. Shaniavskogo za ... akademicheskii god* (Moscow: Tipo-litografiia T-va I. N. Kushnerev i K°, 1909–16).

33. E. W. Clowes, S. D. Kassow and J. L. West (eds), *Between Tsar and People. Educated Society and the Quest for Public Identity in Late Imperial Russia* (Princeton, NJ: Princeton University Press 1991).

34. Doiarenko, *Iz agronomicheskogo proshlogo*, pp. 125–32.

35. That is evidenced by a letter from the chancellery of the Ministry of Internal Affairs to the society, granting a number of agronomists official authorization to conduct lectures and discussions. RGAÈ, f. 9474, op. 1, d. 220, l. 1. On the work of the society, see the minutes of meetings and the annual reports, Central Historical Archive of the City of Moscow (hereafter: TsIAM), f. 1575, op. 1, d. 1, 3 and 23.

36. Of importance were, for example, the Moscow Agronomists' Conference in 1911, the Congress for Small-Scale Credit and Agrarian Cooperatives in 1912 in Petersburg, the All-Russian Cooperative Congresses in 1908 in Moscow and in 1913 in Kiev, and the All-Russian Agricultural Congress in 1913 in Kiev. On the importance of congresses (*s-ezdy*) in the late tsarist empire, see J. Bradley, *Voluntary Associations in Tsarist Russia. Science, Patriotism, and Civil Society* (Cambridge, MA: Harvard University Press, 2009), chapter 6.

37. Musical notes manuscript, Doiarenko, RGAÈ, f. 9474, op. 1, d. 195, l. 5.

38. Ilya Gerasimov, 'Russians into Peasants? The Politics of Self-Organization and Paradoxes of the Public Modernization Campaign in the Countryside in Late Imperial Russia', *Journal of Modern European History*, 2 (2004), pp. 232–52, on p. 240.

39. *Otchet o sostoianii Moskovskogo Sel'skokhoziaistvennogo Instituta za 1910 g.* (Moscow: Tipo-litografiia T-va I. N. Kushnerev i K°, 1911), p. 17.

40. Annual Report of the Society, 1911, TsIAM, f. 1575, op. 1, d. 3.

41. A. V. Chaianov, 'Neobkhodimost' kooperativnogo izucheniia rynkov', *Zemskii Agronom*, 1 (1915), pp. 12–17, on p. 13.

42. Doiarenko, *Krest'ianskie besedy*, pp. 13–4.

43. Kotsonis, *Making Peasants Backward*, p. 96.

44. J. Harwood, *Europe's Green Revolution and Others Since. The Rise and Fall of Peasant-Friendly Plant Breeding* (London: Routledge 2012), chapter 2; F. D'Onofrio, 'Knowing to Transform. Three Ways for Agricultural Economists to Observe Italy, 1900–40' (PhD Dissertation, Universiteit van Utrecht, 2013), pp. 6 and 50.

45. P. Holquist, *Making War, Forging Revolution. Russia's Continuum of Crisis, 1914–1921* (Cambridge, MA: Harvard University Press, 2002), p. 21.

46. Gerasimov, *Modernism and Public Reform*, pp. 148–53. On the importance of social organizations during the war, see also T. Fallows, 'Politics and War Effort in Russia. The Union of Zemstvos and the Organization of the Food Supply, 1914–1916', *Slavic Review*, 37:1 (1978), pp. 70–90.

47. Glavnyi Komitet Vserossiiskogo Soiuza Gorodov, *Trudy Soveshchaniia po èkonomicheskim voprosam, sviazannym s dorogoviznoi i snabzheniem armii, Moskva*

11–13 iiulia 1915 goda (Moscow: 1915), p. 299.

48. A. Stanziani, 'Spécialistes, bureaucrates et paysans. Les approvisionnements agricoles pendant la Première Guerre Mondiale, 1914–1917', *Cahiers du Monde Russe*, 36:1–2 (1995), pp. 71–94.

49. Liga agrarnykh reform, *Organy zemel'noi reformy. Zemel'nye komitety i Liga agrarnykh reform* (Moscow: Universal'naia biblioteka, 1917).

50. 'Sostav Vremennogo Soveta Glavogo Zemel'nogo Komiteta', *Izvestiia Glavnogo Zemel'nogo Komiteta*, 1 (15 July 1917), p. 19.

51. Chaianov, *Chto takoe agrarnyi vopros,*p. 9.

52. D. K. Rowney, *Transition to Technocracy. The Structural Origins of the Soviet Administrative State* (Ithaca, NY: Cornell University Press, 1989), chapter 1.

53. S. Finkel, *On the Ideological Front. The Russian Intelligentsia and the Making of the Soviet Public Sphere* (New Haven, CT: Yale University Press, 2007).

54. On the Narkomzem during the 1920s, see J. W. Heinzen, *Inventing a Soviet Countryside. State Power and the Transformation of Rural Russia, 1917–1929* (Pittsburgh, PA: University of Pittsburgh Press, 2004); M. Wehner, *Bauernpolitik im proletarischen Staat. Die Bauernfrage als zentrales Problem der sowjetischen Innenpolitik 1921 – 1928* (Cologne: Böhlau, 1998).

55. Minutes, Office for the Protection of Agricultural Experimentation, 1918, RGAÈ, f. 478, op. 5, d. 92, l. 102, and following. See also Elina, *Ot tsarskikh sadov,* vol. 2, chapters 4 and 5. A. I. Ugrimov, the head of the Moscow Agricultural Society, became a member of the GOÈLRO-commission in 1920. A. I. Ugrimov, 'Moi put' i rabota v GOÈLRO', in *Sdelaem Rossiiu èlektricheskoi. Sbornik vospominanii uchastnikov Komissii GOÈLRO i stroitelei pervykh èlektrostantsii* (Moscow, Leningrad: Gosènergoizdat, 1961), pp. 83–93.

56. Thus, the Main Land Committee, after the Bolshevik's 'Decree On Land', immediately called for avoiding a redistribution of land; instead, it recommended waiting until the organization of a democratic parliament. 'Ot Glavnogo Zemel'nogo Komiteta', *Vlast' Naroda*, 156 (8 November 1917.), p. 2.

57. Gerasimov underscores the role of the Soviet system of incentives; see his *Modernism and Public Reform*, p. 194.

58. Confirmation of the Status of Doiarenko at the Petrovka Academy, 25 September 1918, RGAÈ, f. 9474, op. 1, d. 217, l. 6–7.

59. Correspondence between the Narkomzem and the Higher Agricultural Courses in Saratov / the Saratov Agricultural Institute, 1918, RGAÈ, f. 478, op. 1, d. 133.

60. Two other prominent examples are A. V. Chaianov's Institute for Agricultural Economics and Politics and N. D. Kondrat'ev's Conjuncture Institute in Moscow. The staff of both institutes had close regular contact with the top executive echelon of various people's commissariats. Commissioned by the Narkomfin, Kondrat'ev's institute carried out studies on the Soviet economy and its development. It was also repeatedly requested by various functionaries in the party and state apparatus to produce expert reports. In 1925, for example, A. I. Rykov, L. D. Trotskii, Ia. A. Iakovlev and G. E. Zinov'ev were among those who turned to the institute. See Record of the Conjuncture Institute, 15 April 1925, RGAÈ, f. 769, op. 1, d. 21, l. 5–21.

61. Among the associates there were N. D. Kondrat'ev, N. P. Makarov, A. V. Chaianov, A. N. Chelintsev, A. V. Teitel and A. G. Doiarenko. A. A. Rybnikov and N. P. Oganovskii also worked regularly for Narkomzem's planning commission *Zemplan*. See the Overview of the Tasks of the Specialists and Scientific Staff Workers in *Zemplan* for 1924–5, RGAÈ f. 478, op. 2, d. 286, l. 1–5, and the minutes of the meeting of the *Zemplan*

steering committee, 1923, RGAĖ f. 478, op. 2, d. 184.

62. 'Plan deiatel'nosti Narkomzema', *Sel'skokhoziaistvennaia Zhizn'*, 2:2 (1922), p. 1; General Plan for the Development of Agriculture, RGAĖ, f. 478, op. 2, d. 156, l. 1–2.

63. P. Mesiatsev, 'Blizhaishchie puti russkoi agronomii i zadachi sel'skokhoziaistvennogo obrazovaniia', *Sel'skokhoziaistvennaia Zhizn'*, 3:7 (1923), pp. 2–4, on p. 3.

64. Statement by the head of the People's Commissariat for Agriculture, A. P. Smirnov, regarding the scientists in his agency. Letter, Smirnov to the Workers' and Peasants' Inspectorate, 10 October 1924, RGAĖ, f. 478, op. 1, d. 1534, l. 3.

65. *XV S-ezd Vsesoiuznoi Kommunisticheskoi Partii – (b). Stenograficheskii otchet* (Moscow, Leningrad: Gosudarstvennoe izdatel'stvo, 1928), p. 1059.

66. Thus, the party functionaries A. P. Smirnov, A. I. Sviderskii and I. A. Teodorovich, who up until that time had supported the experts, were stripped of their leading positions in the Narkomzem in early 1928; on this, see Heinzen, *Inventing a Countryside*, pp. 195–6; Wehner, *Bauernpolitik im proletarischen Staat*, pp. 374–82.

67. An article by G. E. Zinov'ev marked the beginning of the press campaign: G. Zinov'ev, 'Manifest kulatskoi partii', *Bol'shevik*, 13 (1927), pp. 33–47.

68. Kondrat'ev's Conjuncture Institute was shut down in 1928. V. Barnett, 'A Long Wave Goodbye. Kondrat'ev and the Conjuncture Institute, 1920–28', *Europe-Asia Studies*, 47:3 (1995), pp. 413–41, on p. 441. Chaianov's Institute for Agrarian Economy and Policy fell victim to a reorganization of the Timiriazev Academy. S. Solomon Gross, *The Soviet Agrarian Debate. A Controversy in Social Science, 1923–1929* (Boulder, CO: Westview Press, 1977), pp. 156–7.

69. Along with A. V. Chaianov, N. D. Kondrat'ev and N. P. Makarov ranked among the leading agricultural economists in the Soviet Union in the 1920s. Letter, Ia. A. Iakovlev to Council of the People's Commissariats, 16 April 1929, RGAĖ, f. 8390, op. 4, d. 4, l. 1.

70. For the most accurate documentation of the intrigue, see M. L. Galas, *Sud'ba i tvorchestvo russkikh ėkonomistov-agrarnikov i obshchestvenno-politicheskikh deiatelei A. N. Chelintseva i N. P. Makarova* (Moscow: Akson, 2007), chapter 4.

71. See S. Fitzpatrick, 'Stalin and the Making of a New Elite, 1928–1939', *Slavic Review*, 38:3 (1979), pp. 377–402; K. E. Bailes, *Technology and Society Under Lenin and Stalin. Origins of the Soviet Technical Intelligentsia, 1917–1941* (Princeton, NJ: Princeton University Press, 1978); S. Schattenberg, *Stalins Ingenieure. Lebenswelten zwischen Technik und Terror in den 1930er Jahren* (Munich: Oldenbourg, 2002).

72. A. N. Chelintsev was released from prison in 1932. S. L. Maslov returned from banishment in 1933. L. N. Litoshenko and A. A. Rybnikov were released from prison in 1932. Litoshenko joined the staff of the Institute for Metabolism and Endocrine Diseases, while Rybnikov found a position in 1933 in the All-Union Linen Institute. N. P. Makarov and A. V. Chaianov were given an early release from prison and then banished. From 1935 onward, Makarov worked in a grain producing state-owned company (*sovkhoz*) in the Voronezh region. At the end of 1935 Chaianov became an adviser to the Kazakh People's Commissariat for Agriculture. For a more detailed outline of their professional biographies see my *Als das Dorf noch Zukunft war* (Cologne: Böhlau, 2014), chapter 4.2.

73. 'Zhertvy politicheskogo terrora v SSSR', at http://lists.memo.ru (accessed 31 January 2013).

74. On the various stations in Doiarenko's biography, see Kurenyshev, *On slyshal muzyku polei*, pp. 71–80.

75. Excerpts from letter by A. G. Doiarenko to his daughter E. A. Doiarenko, 1935, RGAĖ, f. 9474, op. 1, d. 254, l. 105–19.

76. Kurenyshev, *On slyshal muzyku polei*, pp. 72–3.

77. Letters, Institute to Doiarenko, May and August 1947, RGAĖ, f. 9474, op. 1, d. 223, l. 8–9.

78. V. R. Vil'iams himself had already passed away in 1939.

79. Iu. F. Kurdiukov, 'Saratovskii period tvorcheskoi deiatel'nosti A. G. Doiarenko', *Agrarnyi vestnik Iugo-Vostoka*, 2:5 (2010), pp. 54–8, on p. 56.

80. Kurenyshev, *On slyshal muzyku polei*, pp. 79–80.

81. 'Rech' tovarishcha N. Khrushcheva na soveshchanii rabotnikov sel'skogo khoziaistva Sibiri 26 noiabria 1961 g. v gorode Novosibirske', *Pravda*, 29 November 1961, p. 1.

82. F. Z. Uzarov, Memories of his Studies under A. G. Doiarenko in the 1920s, 1964, RGAĖ, f. 9474, op. 1, d. 255, l. 75.

83. N. V. Orlovskii, Memories of A. G. Doiarenko on the Occasion of his Ninetieth Birthday, 1964, RGAĖ, f. 9474, op. 1, d. 255, l. 6.

84. Hoffmann, *Cultivating the Masses*; D. L. Hoffmann and Y. Kotsonis (eds), *Russian Modernity. Politics, Knowledge, Practices* (Houndsmills: Macmillan Press; New York, NY: St. Martin's Press, 2000); U. Herbert, 'Europe in High Modernity. Reflections on a Theory of the 20th Century', *Journal of Modern European History*, 5:1 (2007), pp. 5–21. For a detailed account of the discussion, see M. David-Fox, 'Multiple Modernities vs. Neo-Traditionalism. On Recent Debates in Russian and Soviet History', *Jahrbücher für Geschichte Osteuropas*, 54:4 (2006), pp. 535–55.

85. J. C. Scott, *Seeing Like a State. How Certain Schemes to Improve the Human Condition Have Failed* (New Haven, CT: Yale University Press, 1998), chapter 6.

86. This interpretation referred especially to the theory of peasant household economy. The discussions of the 1920s have been comprehensively studied by Solomon Gross, *Agrarian Debate*, and T. Cox, *Peasants, Class, and Capitalism. The Rural Research of L. N. Kritsman and his School* (Oxford: Clarendon Press 1986). For a neomarxist criticism of the 'agrarian myth' see T. Brass, 'The Agrarian Myth, the 'New' Populism and the 'New' Right', *Economic and Political Weekly*, 32:4 (1997), pp. 27–42.

7 The Rise of the Scientist-Diplomat within British Atomic Energy, 1945–55

1. A. Brown, *The Neutron and the Bomb: A Biography of Sir James Chadwick* (Oxford: Oxford University Press, 1997), pp. 260–78.

2. J. Krige, 'Britain and the European Laboratory Project: 1951–mid-1952' and 'Britain and the European Laboratory Project: mid-1952–December 1953' in A. Hermann et al. (eds), *History of CERN*, 3 vols (Amsterdam: North Holland, 1987), vol. 1, pp. 431–74 and 475–519.

3. S. Lee (ed.), *Sir Rudolf Peierls: Selected Private and Scientific Correspondence*, 2 vols (London: World Scientific, 2007-2009); Peter Hore, *Patrick Blackett: Sailor, Scientist and Socialist* (London: Frank Cass, 2003).

4. S. Jasanoff, *The Fifth Branch: Science Advisers as Policymakers* (Cambridge, MA: Harvard University Press, 1990), pp. 15–7.

5. J. Cockcroft, 'The Development and Future of Atomic Energy', *Bulletin of the Atomic Scientists*, 6:2 (1950), pp. 325–31, on p. 325.

6. G. Hartcup, *Cockcroft and the Atom* (Bristol: Adam Hilger, 1984), p. 121.

7. Quebec Agreement, 19 August 1943, at http://www.atomicarchive.com/Docs/Man-

hattanProject/Quebec.shtml [accessed 1 October 2012].

8. A. Fort, *Prof: The Life of Frederick Lindemann* (London: Jonathan Cape, 2003), p. 218.

9. H. Massey, 'Atomic Energy and the Development of Large Teams and Organizations', *Proceedings of the Royal Society of London, Series A: Mathematical and Physical Sciences*, 342:1631 (1975), pp. 491–7, on p. 495.

10. J. Rotblat, 'Leaving the Bomb Project', *Bulletin of the Atomic Scientists*, 41:7 (1985), pp. 16–9, on p. 18.

11. Resolution Adopted at the Council Meeting of the Association of Scientific Workers, 19 November, 1938, Royal Society Archives London, Patrick Blackett Papers (hereafter RS PB), 5/1/E.3.

12. P. Gummett, *Scientists in Whitehall* (Manchester: Manchester University Press, 1980), pp. 101–2.

13. C. Laucht, *Elemental Germans: Klaus Fuchs, Rudolf Peierls and the Making of British Nuclear Culture 1939–59* (London: Macmillan, 2012), pp. 131-2.

14. The Russell-Einstein Manifesto, issued in London, 9 July 1955 at http://scarc.library.oregonstate.edu/coll/pauling/peace/papers/peace6.007.5-01.html [accessed 13 January 2015].

15. 1945 Labour Party Election Manifesto, at http://www.labour-party.org.uk/manifestos/1945/1945-labour-manifesto.shtml [accessed 1 October 2012].

16. Washington Declaration, 15 November 1945, quoted in Margaret Gowing, *Independence and Deterrence: Britain and Atomic Energy, 1945–52*, 2 vols (London: Macmillan, 1974), vol. 1, pp. 82-4.

17. Groves-Anderson Memorandum, 16 November 1945, quoted in Gowing, *Independence and Deterrence*, vol. 1, p. 85; J. Helmreich, *Gathering Rare Ores: The Diplomacy of Uranium Acquisition* (Princeton, NJ: Princeton University Press, 1986), p. 107.

18. P. Hennessy, *Cabinets and the Bomb* (Oxford: Oxford University Press, 2007), p. 7.

19. Hansard House of Lords Debate, 5 July 1951, vol. 172, cols 670–9, at http://hansard.millbanksystems.com/lords/1951/jul/05/atomic-energy [accessed 3 February 2015]; Brown, *The Neutron and the Bomb*, pp. 359–60.

20. B.R. Mitchell, *2nd Abstract of British Historical Statistics* (Cambridge: Cambridge University Press, 1971), pp. 66-8.

21. D. Painter, *Oil and the American Century: The Political Economy of U.S. Foreign Oil Policy 1941–1954* (Baltimore, MD: Johns Hopkins University Press, 1986), pp. 153–65.

22. M. Gowing, *Independence and Deterrence: Britain and Atomic Energy, 1945–52*, 2 vols (London: Macmillan, 1974), vol. 2, p. 283 and pp. 302–12.

23. M. J. Nye, *Blackett: Physics, War and Politics in the Twentieth Century* (Cambridge, MA: Harvard University Press, 2004), p. 86.

24. Brown, *The Neutron and the Bomb*, pp. 328–9.

25. Memorandum to James Chadwick and John Cockcroft by Montreal Scientists, 30 August 1945, RS PB, 4/11/D.185

26. Letter from Mark Oliphant to Patrick Blackett, 22 January 1946, RS PB, 4/11/D.192.

27. Brown, *The Neutron and the Bomb*, pp. 260–78.

28. Minutes of a Cabinet Meeting, 17 January 1946, National Archives, Kew Gardens, Surrey, Cabinet Papers (hereafter NA CAB), 128/5.

29. J. Cockcroft, 'The Scientific Work of the Atomic Energy Research Establishment', *Proceedings of the Royal Society of London, Series A: Mathematical and Physical Sciences*, 211:1105 (1952), pp. 155–68, on p. 155.

30. R. Pocock, *Nuclear Power: Its Development in the United Kingdom* (Old Woking:

Unwin Brothers, 1977), p. 8; Gowing, *Independence and Deterrence*, vol. 2, pp. 442–5.

31. Helmreich, *Gathering Rare Ores*, pp. 15–41, p. 68, pp. 72–96 and 102-3.
32. United States Atomic Energy Commission, 'Fifth Semi-Annual Report, Part II', *Bulletin of the Atomic Scientists*, 5:4 (1949), pp. 114–125, on p. 121.
33. Gowing, *Independence and Deterrence*, vol. 1, pp. 245–6.
34. Gowing, *Independence and Deterrence*, vol. 2, p. 85.
35. Gowing, *Independence and Deterrence*, vol. 1, pp. 38–43 and Letter from Oliver Franks to John Cockcroft, 9 November 1945, Churchill College Cambridge Archives, John Cockcroft Papers (hereafter CC CKFT), 23/2.
36. Gowing, *Independence and Deterrence*, vol. 2, pp. 69–70 and p. 206.
37. United Kingdom Atomic Energy Act 1946.
38. E. Shils, 'British Atomic Energy Act Debate', *Bulletin of the Atomic Scientists*, 3:2 (1947), pp. 52–4.
39. 'Atomic Energy: Letter to the Editor', *Times*, 5 November 1946, p. 5.
40. Memorandum from John Cockcroft to R.E. France, 6 September 1951, National Archives, Kew Gardens, Surrey, Records of the United Kingdom Atomic Energy Authority and its Predecessors (hereafter NA AB), 27/6.
41. Committee Organization for Dealing with Atomic Energy Matters, Note by the Secretary of the Cabinet, 10 February, 1948, NA CAB, 129/24.
42. Letter from Clement Attlee to Patrick Blackett, 6 January 1948, RS PB, 4/11/D.184.
43. Cockcroft, 'The Development and Future of Atomic Energy', pp. 325–31, on p. 325.
44. John Cockcroft, 'Future of Atomic Energy', *Scientific Monthly*, 82:3 (1956), pp. 136–41, on p. 138.
45. Nye, *Blackett: Physics, War and Politics in the Twentieth Century*, pp. 158–68.
46. Nye, *Blackett: Physics, War and Politics in the Twentieth Century*, pp. 88–9.
47. P. Blackett, *Military and Political Consequences of Atomic Energy* (London: Turnstile Press, 1948), p. 188.
48. Letter from Lord Cherwell to John Cockcroft, 7 March 1952, CC CKFT, 23/2.
49. Letter from Lord Cherwell to Patrick Blackett, 18 May 1946, RS PB, 4/11/D.197.
50. Letter from Rudolf Peierls to James Chadwick, 3 May 1946, quoted in S. Lee (ed.), *Sir Rudolf Peierls: Selected Private and Scientific Correspondence*, 2 vols (London: World Scientific, 2007–9), vol. 2, pp. 48–54.
51. Personal Minute from Winston Churchill to Duncan Sandys, 15 November 1951, NA AB, 19/57.
52. Atomic Energy Organization: Transfer from Ministry of Supply to a National Corporation, Memorandum from the Paymaster-General, 30 September 1952, NA CAB, 129/55.
53. Lord Walter Citrine, *Two Careers* (London: Hutchinson, 1967), pp. 302–5.
54. Atomic Energy Organization: Transfer from Ministry of Supply to a National Corporation, Memorandum from the Paymaster-General, 30 September 1952, NA CAB, 129/55.
55. Atomic Energy Organization, Memorandum from the Minister of Supply, 25 October 1952, NA CAB, 129/56.
56. Atomic Energy Organization: Memorandum by the Secretary of State for the Colonies, 24 October 1952, NA CAB, 129/56.
57. The Athenaeum Club is a London gentleman's club popular with officials in Whitehall.
58. Conclusions of a Meeting of the Cabinet, 6 November 1952, NA CAB, 128/25, pp. 77–9.
59. Pocock, *Nuclear Power: Its Development in the United Kingdom*, pp. 25–6.
60. Report of the Committee on the Future Organization of the Atomic Energy Project, 23 July 1953, NA CAB, 129/62, pp. 19–20.

61. United Kingdom Atomic Energy Authority Act, 1954.

62. Gowing, *Independence and Deterrence*, vol. 2, p. 58.

63. Memorandum from Donald Perrott to John Cockcroft, 12 January 1954, CC CKFT, 23/2.

64. Memorandum from Christopher Hinton to Edwin Plowden, 28 March, 1955, NA AB, 19/13.

65. Frederick Furneaux Smith, *The Prof in Two Worlds: The Official Life of Professor F. A. Lindemann, Viscount Cherwell* (London: Collins, 1961), p. 315.

66. Hartcup, *Cockcroft and the Atom*, p. 191.

67. Sir Frederick Brundrett, 'Government and Science', *Public Administration*, 3:3 (1956), pp. 245–56, on p. 250.

68. Brown, *The Neutron and the Bomb*, pp. 333–7.

69. Exchanges between Polish Minister for Atomic Energy and John Cockcroft, May–June 1957 and Lecture Given to the Polish Academy of Science, 22 March 1957, NA AB, 27/14.

70. A Programme of Nuclear Power, White Paper, February 1955, NA CAB, 129/73.

71. J. Krige, 'Atoms for Peace, Scientific Internationalism and Scientific Intelligence', *Osiris*, 21:1 (2006), pp. 161–181, on p. 175, n. 64. NA AB, 6/1663 contains details, including programmes and timetables, of the visits by foreign scientists to Harwell.

72. Letter from Edwin Plowden to Harold Macmillan, 16 July 1957, National Archives, Kew Gardens, Surrey, Records of the Prime Minister's Office, NA PREM, 11/2848. Additionally, NA AB, 41/349 contains reports from meetings of the UKAEA's Euratom Steering Committee, chaired by Penney.

73. D. Pestre, 'Regimes of Knowledge Production in Society: Towards a More Political and Social Reading', *Minerva*, 41:3 (2003), pp. 245–61, on p. 260.

74. M. Gowing, 'Lord Hinton of Bankside', *Biographical Memoirs of Fellows of the Royal Society*, 36 (1990), pp. 219–39, on p. 226.

75. Speech Opening the Science Debate at the Party's Annual Conference, Scarborough, 1963, in H. Wilson, *Purpose in Politics: Selected Speeches* (London: Weidenfeld and Nicolson, 1964), p. 27, quoted in D. Edgerton, 'The "White Heat" Revisited: The British Government and Technology in the 1960s', *Twentieth Century British History*, 7:1 (1996), p. 56.

8 The Reform Technocrats: Strategists of the Swedish Welfare State, 1930–60

1. A classic in this respect is M. W. Childs, *Sweden: The Middle Way* (New Haven, CT: Yale University Press, 1936), which first helped to elevate Sweden to iconic status. See, for instance, J. Louge, 'The Swedish Model: Visions of Sweden in American Politics and Political Science', *The Swedish-American Historical Quarterly*, 50:3 (1999), pp. 162–72; K. Musiał, *Roots of the Scandinavian Model: Images of Progress in the Era of Modernization* (Baden-Baden: Nomos Verlagsgesellschaft, 2002); A. Ruth, 'The Second New Nation: The Mythology of Modern Sweden', *Dædalus*, 113:2 (1984), pp. 53–96, on pp. 64–5.

2. Admittedly, a host of studies – especially in the history of science and technology – have considered the role of experts and expert knowledge in modern Swedish society. However, these particular histories seldom find their way into the grand narrative of the rise of the welfare state. And even if they do, engineers, planners, scientists and other professionals are rarely assigned a crucial role in the formation of Sweden's welfare

society. Recent titles that match this description include K. Åmark, *Hundra år av välfärdspolitik: Välfärdsstatens framväxt i Norge och Sverige* (Umeå: Boréa, 2005); Y. Hirdman, U. Lundberg and J. Björkman, *Sveriges historia, 1920–1965* (Stockholm: Norstedt, 2012); S. E. O. Hort, *Social Policy, Welfare State, and Civil Society in Sweden: Vol. 1, History, Policies, and Institutions, 1884–1988* (Lund: Arkiv, 2014); F. Sejersted, *The Age of Social Democracy: Norway and Sweden in the Twentieth Century* (Princeton, NJ: Princeton University Press, 2011).

3. P. Lundin, N. Stenlås and J. Gribbe (eds), *Science for Welfare and Warfare: Technology and State Initiative in Cold War Sweden* (Sagamore Beach, MA: Science History Publications, 2010).

4. P. Lundin and N. Stenlås, 'Technology, State Initiative and National Myths in Cold War Sweden: An Introduction', in Lundin, Stenlås and Gribbe (eds), *Science for Welfare and Warfare*, pp. 1–34.

5. C. Maier, 'Society as a Factory', in C. Maier (ed.), *In Search of Stability: Explorations in Historical Political Economy* (Cambridge: Cambridge University Press, 1987), pp. 19–69, on p. 28.

6. T. Judt, *Postwar: A History of Europe since 1945* (London: Heinemann, 2005).

7. The central role of Social Democracy and the Social Democratic ideology in shaping Swedish post-war society is maintained in V. Bergström, 'Party Program and Economic Policy: The Social Democrats in Government', in K. Misgeld, K. Molin and K. Åmark (eds), *Creating Social Democracy: A Century of the Social Democratic Labor Party in Sweden* (University Park, PA: Pennsylvania State University Press, 1992), pp. 131–73; G. Esping-Andersen, 'The Making of a Social Democratic Welfare State', in Misgeld, Molin and Åmark (eds), *Creating Social Democracy*, pp. 35–66; G. Therborn, 'A Unique Chapter in the History of Democracy: The Social Democrats in Sweden', in Misgeld, Molin and Åmark (eds), *Creating Social Democracy*, pp. 1–34; T. Tilton, 'The Role of Ideology in Social Democratic Politics', in Misgeld, Molin and Åmark (eds), *Creating Social Democracy*, pp. 409–27.

8. P. A. Swenson, *Capitalists Against Markets: The Making of Labor Markets and Welfare States in the United States and Sweden* (New York, NY: Oxford University Press, 2002); E. Uddhammar, *Partierna och den stora staten: En analys av statsteorier och svensk politik under 1900-talet* (Stockholm: City University Press, 1993).

9. W. Korpi, *The Democratic Class Struggle* (London: Routledge & Kegan Paul, 1983).

10. Even if this dominant perspective is being challenged, it still shapes the historiography of the welfare state to a considerable extent. U. Lundberg and M. Tydén, 'In Search of the Swedish Model: Contested Historiography', in H. Mattsson and S.-O. Wallenstein (eds), *Swedish Modernism: Architecture, Consumption and the Welfare State* (London: Black Dog Publishing, 2010), pp. 36–49.

11. Y. Hirdman, *Att lägga livet tillrätta: Studier i svensk folkhemspolitik*, 2nd edn (Stockholm: Carlsson, 2000); G. B. Nilsson, 'Den sociala ingenjörskonstens problematik', in P. Thullberg and K. Östberg (eds), *Den svenska modellen* (Lund: Studentlitteratur, 1994), pp. 161–79; U. Sandström, *Arkitektur och social ingenjörskonst: Studier i svensk arkitektur och bostadsforskning* (Linköping: Linköpings universitet, 1989).

12. A recent example that discusses expertise and the rise of the welfare state in this rather narrow sense is Å. Lundqvist and K. Petersen (eds), *In Experts We Trust: Knowledge, Politics and Bureaucracy in Nordic Welfare States* (Odense: University Press of Southern Denmark, 2010).

13. K. Mannheim, *Ideology and Utopia*, trans. L. Wirth and E. Shils (San Diego, CA:

Harcourt, Inc., 1985), p. 155.
14. R. Eyerman, *Between Culture and Politics: Intellectuals in Modern Society* (Cambridge: Polity Press, 1994), pp. 126–8.
15. R. Slagstad, *De nasjonale strateger* (Oslo: Pax, 1998), p. 168.
16. Recently, Swedish historians of ideas have acknowledged the social engineer as a historical category, and stressed that as such it differs substantially from the welfare studies' analytical category. As a historical category, however, the social engineers are a group of limited significance for the processes and historical period we are interested in. H. Björck, *Folkhemsbyggare* (Stockholm: Atlantis, 2008), pp. 53–90; U. Larsson, 'Socialingenjören: Hjalmar Cederström och kampen mot sjukdom och fattigdom', in S. Widmalm (ed.), *Vetenskapsbärarna: Naturvetenskapen i det svenska samhället, 1880–1950* (Hedemora: Gidlunds förlag, 1999), pp. 273–316; D. Östlund, *Det sociala kriget och kapitalets ansvar: Social ingenjörskonst mellan affärsintresse och samhällsreform i USA och Sverige, 1899–1914* (Stockholm: Stockholms universitet, 2003).
17. The number rose from 21 per cent in 1950 to 44 per cent in 1965. G. C.-O. Claesson, *Statens ostyriga utredande: Betänkande om kommittéväsendet* (Stockholm: Studieförbundet näringsliv och samhälle, 1972), pp. 43–4.
18. J. Johansson, *Det statliga kommittéväsendet: Kunskap, kontroll, konsensus* (Stockholm: Stockholms universitet, 1992); H. Meijer, *Kommittépolitik och kommittéarbete: Det statliga kommittéväsendets utvecklingslinjer, 1905–1954, samt nuvarande funktion och arbetsformer* (Lund: Gleerup, 1956).
19. Claesson, *Statens ostyriga utredande;* P. Wisselgren, 'Reforming the Science-Policy Boundary: The Myrdals and the Swedish Tradition of Governmental Commissions', in S. Eliaeson and R. Kalleberg (eds), *Academics as Public Intellectuals* (Newcastle: Cambridge Scholars Publishing, 2008), pp. 173–95.
20. G. Myrdal, 'Bostadspolitiska preludier', in G. Myrdal et al. (eds), *Bostadspolitik och samhällsplanering: Nio uppsatser* (Stockholm: Tiden, 1968), pp. 9–14; G. Myrdal, *Hur styrs landet?* (Stockholm: Rabén & Sjögren, 1982), p. 144 and pp. 190–2.
21. T. Forser, *'Jag har speglat århundradet' – En bok om Per Nyström: Historikern, publicisten, ämbetsmannen* (Stockholm: Rabén Prisma, 1996), p. 152.
22. P. Lundin, *Bilsamhället: Ideologi, expertis och regelskapande i efterkrigstidens Sverige* (Stockholm: Stockholmia förlag, 2008); S. Wallander, 'Vad har vi gjort?' in G. Myrdal et al. (eds), *Bostadspolitik och samhällsplanering*, pp. 24–6.
23. A. Clason, *Plantänkande och planering: En studie i 1940-talets samhällsplaneringsidéer och initiativ* (Gävle: Statens institut för byggnadsforskning, 1982).
24. Lundin, *Bilsamhället;* E. Rudberg, *Uno Åhrén: En föregångsman inom 1900-talets arkitektur och samhällsplanering* (Stockholm: Statens råd för byggnadsforskning, 1981).
25. Yet another example is the married couple Axel and Signe Höjer. Both were influential in shaping health care and social reforms. A. Berg, *Den gränslösa hälsan: Signe och Axel Höjer, folkhälsan och expertisen* (Uppsala: Uppsala universitet, 2009).
26. S. Vidén, 'Rekordårens bostadsbyggande', in Thomas Hall (ed.), *Rekordåren: En epok i svenskt bostadsbyggande* (Karlskrona: Boverket, 1999), p. 33.
27. Further examples include the jurist Carl-Henrik Nordlander, who had a key role in the 1942 Agricultural Committee and went on to become state secretary at the Ministry of Agriculture, and later also director general of the reformed Royal Board of Agriculture; as well as the electrical engineer Åke Rusck, who occupied a central function in the 1943 Electric Power Committee, which he left to become director general of the reformed Royal Waterfall Board. H. Jörgensen, 'Neutrality and National Preparedness:

State-Led Agricultural Rationalization in Cold War Sweden', in Lundin, Stenlås and Gribbe (eds), *Science for Welfare and Warfare*, pp. 173–94; Lundin and Stenlås, 'Technology, State Initiative and National Myths in Cold War Sweden'.

28. Lundin, *Bilsamhället*, pp. 61–2.
29. M. Andersson, *Politik och stadsbyggande: Modernismen och lagstiftningen* (Stockholm: Stockholms Universitet, 2009), p. 152, pp. 181–2, pp. 192–3.
30. T. Nybom, 'The Socialization of Science: Technical Research and the Natural Sciences in Swedish Research Policy in the 1930s and 1940s', in S. Lindqvist (ed.), *Center on the Periphery: Historical Aspects of 20th-Century Swedish Physics* (Canton, MA: Science History Publications, 1993), pp. 164–78, on p. 168; H. Weinberger, *Nätverksentreprenören: En historia om teknisk forskning och industriellt utvecklingsarbete från den Malmska utredningen till Styrelsen för teknisk utveckling* (Stockholm: Kungl. Tekniska högskolan, 1997), p. 96.
31. Nybom, 'The Socialization of Science', p. 169.
32. See T. Petersson, 'Private and Public Interests in the Development of the Early Swedish Computer Industry: Facit, Saab and the Struggle for National Dominance', in Lundin, Stenlås and Gribbe (eds), *Science for Welfare and Warfare*, pp. 109–30; G. Holmberg, 'Public Health, National Security and Food Technology in the Cold War: The Swedish Institute for Food Preservation Research', in Lundin, Stenlås and Gribbe (eds), *Science for Welfare and Warfare*, pp. 195–212.
33. O. Månsson, *Industriell beredskap: Om ekonomisk försvarsplanering inför andra världskriget* (Stockholm: LiberFörlag, 1976); H. Weinberger, 'Physics in Uniform: The Swedish Institute of Military Physics, 1939–1945', in Lindqvist (ed.), *Center on the Periphery*, pp. 141–63, on pp. 146–7.
34. W. Agrell, *Vetenskapen i försvarets tjänst: De nya stridsmedlen, försvarsforskningen och kampen om det svenska försvarets struktur* (Lund: Lund University Press, 1989), pp. 103–11; Weinberger, 'Physics in Uniform', pp. 153–6.
35. Agrell, *Vetenskapen i försvarets tjänst*, pp. 134–5; T. Jonter, *Sweden and the Bomb: The Swedish Plans to Acquire Nuclear Weapons, 1945–1972* (Stockholm: SKI, 2001).
36. Sigurd Nauckhoff, chairman of the Royal Swedish Academy of Engineering Sciences, became the first managing director of AB Atomenergi. M. Fjæstad and T. Jonter, 'Between Welfare and Warfare: The Rise and Fall of the "Swedish Line" in Nuclear Engineering', in Lundin, Stenlås and Gribbe (eds), *Science for Welfare and Warfare*, pp. 153–72; S. Lindström, *Hela nationens tacksamhet: Svensk forskningspolitik på atomenergiområdet, 1945–1956* (Stockholm: Stockholms universitet, 1991), pp. 92–3.
37. According to Arthur Engberg's successor, Tage Erlander, primary and lower secondary education were none of his priorities. Sejersted, *The Age of Social Democracy*, p. 267.
38. Sejersted, *The Age of Social Democracy*, p. 269; T. Erlander, *Tage Erlander: 1940–1949* (Stockholm: Tiden, 1973), p. 234; B. Rothstein, *Den socialdemokratiska staten: Reformer och förvaltning inom svensk arbetsmarknads- och skolpolitik* (Lund: Arkiv, 1986), p. 130.
39. See, for instance, S. Marklund, *Skolsverige, 1950–1975: Del 1, 1950 års reformbeslut* (Stockholm: Liber, 1980), pp. 50–8; R. G. Paulston, *Educational Change in Sweden: Planning and Accepting the Comprehensive School Reforms* (New York, NY: Teachers College Press, 1968), p. 68; G. Richardson, *Svensk skolpolitik, 1940–1945* (Stockholm: LiberFörlag, 1978), p. 256.
40. Richardson, *Svensk skolpolitik, 1940–1945*, p. 54.
41. A. Norberg, B. Asker and A. Tjerneld, *Tvåkammarriksdagen, 1867–1970: Ledamöter och valkretsar*, 5 vols (Stockholm: Almqvist & Wiksell International, 1985–1992), vol.

4, pp. 279–80.

42. Richardson, *Svensk skolpolitik*, p. 252.

43. Tage Erlander's diary, 18 October 1945, printed in T. Erlander, *Dagböcker: 1945–1949* (Hedemora: Gidlunds förlag, 2001), p. 23.

44. Tage Erlander's diary, 2 January 1946, printed in Erlander, *Dagböcker*, p. 59.

45. They were Alva Myrdal and Adolf Wallentheim. Marklund, *Skolsverige, 1950–1975: Del 1*, p. 64 and 133.

46. Marklund, *Skolsverige, 1950–1975: Del 1*, pp. 133–4.

47. Marklund, *Skolsverige, 1950–1975: Del 1*, p. 70; G. Richardson, '1950 års en-hetsskolebeslut – en politisk nebulosa', *Statsvetenskaplig Tidskrift*, 70:5 (1967), pp. 369–94.

9 Rationalization Comes to Rome: Expertise in Labour Management at the Third International Congress, 1927

1. E.B. Cooke, 'The Third International Management Congress, Held at Rome, Italy, September 5–9, 1927', *Bulletin of the Taylor Society* (October 1927), pp. 486–90, on p. 488.

2. Cooke, 'The Third International Management Congress', p. 488.

3. T. Broman, 'The Semblance of Transparency: Expertise as a Social Good and an Ideology in Enlightened Societies', in R.E. Kohler and K.M. Olesko (eds), *Clio Meets Science: The Challenges of History* (*Osiris*, 27, 2012), pp. 188–208.

4. See J. Vandendriessche, E. Peeters and K. Wils, 'Introduction: Performing Expertise', in this volume.

5. M. Weber, *The Protestant Ethic and the Spirit of Capitalism* (London: Routledge, 2001); see also M. Foucault, *Discipline and Punish: The Birth of the Prison* (New York, NY: Random House, 1977) and P. S. Gorski, *The Disciplinary Revolution: Calvinism and the Rise of the State in Early Modern Europe* (Chicago, IL: University of Chicago Press, 2003).

6. C. S. Maier, 'Between Taylorism and Technocracy: European Ideologies and the Vision of Industrial Productivity in the 1920s', *Journal of Contemporary History*, 5 (1970), pp. 27–61.

7. R. W. Scheffler, 'The Fate of a Progressive Science: The Harvard Fatigue Laboratory, Athletes, the Science of Work and the Politics of Reform', *Endeavour*, 35 (2011), pp. 48–54; R. Kanigel, *The One Best Way: Frederick Winslow Taylor and the Enigma of Efficiency* (New York, NY: Viking, 1997), pp. 486–503; H. Homburg, *Rationalisierung und Industriearbeit: Arbeitsmacht– Management– Arbeiterschaft im Siemens-Konzern Berlin, 1900–1939* (Berlin: Haude & Spener, 1991).

8. D. Peukert, *The Weimar Republic: The Crisis of Classical Modernity* (New York, NY: Hill and Wang, 1989), p. vii; R. J. Evans, *The Coming of the Third Reich* (New York, NY: Penguin Press, 2004), pp. 109–117; P. Malanima and V. Zamagni, '150 Years of the Italian Economy, 1861–2010', *Journal of Modern Italian Studies*, 15 (2010), pp. 1–20, on p. 6; R. J. B. Bosworth, *Mussolini's Italy: Life Under the Fascist Dictatorship, 1915–1945* (New York, NY: Penguin Press, 2005), pp. 1 and 249–76; P. R. Willson, *The Clockwork Factory: Women and Work in Fascist Italy* (Oxford: Clarendon Press, 1993).

9. My thanks to Niklas Stenlås for providing biographical details for Kaernekull. Cooke, 'The Third International Management Congress', p. 489; Cooke saw Leffingwell and Clark at the meeting, Kaernekull described attending in his personal notes and the published copy of Ascher's presentation indicates that he delivered it by mail.

10. *Atti del III° Congresso Internazionale di Organizzazione Scientifica del Lavoro* (Rome: L'Universerale, 1927).

11. W. Clark, 'The Technic of Installation of Scientific Management', *Atti del III° Congresso Internazionale*, pp. 449–57.

12. W. Clark, *The Gantt Chart* (New York, NY: The Ronald Press Company, 1922); W. Clark, *Foremanship: The Standard Course of the United YMCA Schools* (New York, NY: Association Press, 1921).

13. W. Pidgajetzky, 'Problem der physiologischen Rationalisierung der Frauen-Arbeit auf den Zuckerrüben-Feldern', *Atti del III° Congress Internazionale*, pp. 415–20.

14. E. Rognon, 'L'organisation du travail à la 'Société des Transports en Commun de la Région Parisienne', *Atti del III° Congresso Internazionale*, pp. 1027–39.

15. L. Ascher, 'Der Einfluss technischer Verbesserungen auf die Gesundheit des Menschen, insbesondere des Arbeiters', *Atti del III° Congresso Internazionale*, pp. 563–70.

16. J. K. Alexander, *The Mantra of Efficiency: From Waterwheel to Social Control* (Baltimore, MD: Johns Hopkins University Press, 2008), pp. 101–25.

17. W. H. Leffingwell, 'The Application of Scientific Management to the Office', *Atti del III° Congresso Internazionale*, pp. 5–20.

18. W. Leffingwell, *Making the Office Pay: Tested Office Plans, Methods, and Systems that Make for Better Results from Everyday Routine* (New York, NY: A.W. Shaw Company, 1918).

19. J. K. Alexander, 'Efficiency and Pathology: Mechanical Discipline and Efficient Worker Seating in Germany, 1929–1932', *Technology and Culture*, 47 (2006), pp. 286–310.

20. O. Kaernekull, 'Experiences in the Adoption of Rational Industrial Management in Sweden', *Atti del III° Congresso Internazionale*, pp. 292–4, on p. 292.

10 Scientific Expertise in Child Protection Policies and Juvenile Justice Practices in Twentieth-Century Belgium

1. See J. Vandendriessche, K. Wils and E. Peeters, 'Introduction: Performing Expertise', in this volume.

2. C. Debuyst, F. Digneffe and A. Pires, *Histoire des savoirs sur le crime et la peine. Tome 2: La rationalité pénale et la naissance de la criminologie* (Montréal and Brussels: Presses de l'Université de Montréal, Presses de l'Université d'Ottawa, De Boeck Université, 1998), pp. 287–97. The doctrine of 'social defence' gained immediate and durable success via the support of the International Union of Criminal Law, founded in 1889. A. Prins, *La défense sociale et les transformations du droit pénal* (Brussels: Misch et Thron, 1910); F. Tulkens, 'Adolphe Prins et la défense sociale', in F. Tulkens (ed.), *Généalogie de la défense sociale en Belgique (1880–1914)* (Brussels: Story-Scientia, 1988).

3. D. Niget, 'L'enfance irrégulière et le gouvernement du risque', in D. Niget and M. Petitclerc (eds), *Pour une histoire du risque: Québec, France, Belgique* (Québec and Rennes: Presses de l'Université du Québec, Presses universitaires de Rennes, 2012), pp. 297–316.

4. R. De Bont, 'Meten en Verzoenen. Louis Vervaeck en de Belgische criminele antropologie, circa 1900–1940', *Cahiers d'Histoire du Temps Présent*, 9 (2001), pp. 63–104.

5. L. Vervaeck, 'La conception anthropologique du traitement des condamnés. Les réformes du système pénitentiaire qu'elle entraine', *Revue de Droit pénal et de criminologie*, 14:1 (1921), pp. 368–9.

6. P. Vervaeck, 'Délinquance et criminalité de l'enfance', *Revue de Droit pénal et de crimi-*

nologie, 29:2 (1936), pp. 706–21, 894–911, 1087–1112.

7. M. Depaepe, *Zum Wohl des Kindes? Pädologie, pädagogische Psychologie und Experimentelle Pädagogik in Europa und den USA, 1890–1940* (Leuven: Leuven University Press, 1993).

8. The first version of the 'metric intelligence meter' appears in 1905 and it will be reshaped all along the century by numerous psychologists and educators. A. Ohayon, J. Carroy and R. Plas, *Histoire de la psychologie en France: XIXe–XXe siècles* (Paris: La Découverte, 2006), pp. 99–100.

9. M. J. Ratcliff and M. Ruchat, *Les laboratoires de l'esprit: une histoire de la psychologie à Genève, 1892–1965* (Le Mont-sur-Lausanne: LEP, 2006).

10. T. R. Richardson, *The Century of the Child: The Mental Hygiene Movement and Social Policy in the United States and Canada* (Albany, NY: State University of New York Press, 1989).

11. M. Ruchat, *Inventer les arriérés pour créer l'intelligence: l'arriéré scolaire et la classe spéciale: histoire d'un concept et d'une innovation psychologique, 1874–1914* (Bern: Peter Lang, 2003).

12. *Premier congrès international de la protection de l'enfance. Tome 1ᵉʳ: Débats*

13. R. Luaire, *Le rôle de l'initiative privée dans la protection de l'enfance délinquante en France et en Belgique* (Paris: LGDJ, 1936), p. 282.

14. M. Depaepe, F. Simon and A. Van Gorp, 'L'expertise d'Ovide Decroly à l'égard de la délinquance et de la criminalité juvéniles', in A. François, V. Massin and D. Niget (eds), *Violences juvéniles sous expertise(s), XIXe–XXIe siècles. Expertise and Juvenile Violence, 19th–21st Century* (Louvain-la-Neuve: Presses universitaires de Louvain, 2011), pp. 39–64.

15. D. S. Tanenhaus, *Juvenile Justice in the Making* (Oxford and New York, NY: Oxford University Press 2006).

16. J.-F. Crombois, *L'univers de la sociologie en Belgique de 1900 à 1940* (Brussels: Université de Bruxelles, 1998).

17. A. Racine, *Les enfants traduits en justice; étude d'après trois cents dossiers du Tribunal pour enfants de l'arrondissement de Bruxelles* (Liège: G. Thone, 1935).

18. G. Zélis, 'Formation au travail social et mouvement d'éducation ouvrière en Belgique: genèse et organisation des écoles sociales durant l'Entre-deux-guerres', *Vie Sociale*, 2 (2000), pp. 41–62.

19. O. Ihl and M. Kaluszynski, 'Pour une sociologie historique des sciences de gouvernement', *Revue française d'administration publique*, 102:2 (2002), pp. 229–43.

20. A. François, *Guerres et délinquance juvénile. Un demi-siècle de pratiques judiciaires et institutionnelles envers des mineurs en difficulté (1912–1950)* (Brussels and Bruges: La Charte, Die Keure, 2011), p.54; M. De Koster, 'Tot maat van het recht. De vroege ontwikkeling van de wetenschap van het ontspoorde en criminele kind in het Centrale Observatiegesticht in Mol (1913–1941)', in N. Bakker, S. Braster, M. Rietveld-van Wingerden and A. Van Gorp (eds), *Kinderen in gevaar. Jaarboek voor de Geschiedenis van Opvoeding en Onderwijs 2007* (Assen: Van Gorcum, 2007), pp. 94–119, on pp. 100 and 102.

21. Conclusions based on individual case files of boys detained at Mol in years 1916, 1921, 1931 and 1941. M. De Koster, 'Tot maat van het recht', p. 100.

22. François, *Guerres et délinquance juvénile*, pp. 258–61.

23. François, *Guerres et délinquance juvénile*, p. 259.

24. 'Annexe V: Rapport du juge des enfants de Huy, Derriks', *Procès verbal de l'assemblée générale de l'Union des juges des enfants du 6 juillet 1919*, pp. 157–8, cited in François,

Guerres et délinquance juvénile, p. 215.

25. J. Christiaens, *De geboorte van de jeugddelinquent. België, 1830–1930* (Brussels: VUB Press, 1999), p. 316.

26. 'Jaarverslagen / rapports annuels 1930–1989', Mol Observation Centre records (COG Mol 2000), Belgian State Archives, Beveren, annual reports 35–86.

27. M. Rouvroy, 'La clinique psychologique belge de la protection de l'enfance', *Revue belge de Pédagogie*, 15 (1933–4), pp. 143–52, on p. 143.

28. M. Rouvroy, *L'observation pédagogique des enfants de justice* (Brussels: Office de Publicité, 1921), pp. 136–46.

29. Rouvroy, *L'observation pédagogique des enfants de justice*, p. 23; P. Wets, *Enfance coupable et tribunaux pour enfants* (Louvain: Éditions de la société d'études morales, sociales et juridiques, 1937), p. 69.

30. 'Les tests sont appelés à fournir des pistes pour l'observation quotidienne et on les invoque pour vérifier les résultats de cette observation ... ils sont pour nous des moyens d'orientation préalable, des moyens de contrôle et des moyens d'expression lors du diagnostic final', in Rouvroy, *L'observation pédagogique des enfants de justice*, pp.174–5.

31. Rouvroy, *L'observation pédagogique des enfants de justice*, pp. 198–203.

32. A. Racine, *La délinquance des enfants dans les classes aisées* (Brussels: Librairie Falk fils, G. Van Campenhout, 1939), p. 60.

33. De Koster, 'Tot maat van het recht', p. 100.

34. M. Veys, *Le caractère des adolescents débiles mentaux et délinquants* (unpublished doctoral dissertation, Louvain: Université catholique de Louvain, 1934).

35. L. Leroux, *Éducation dans la confiance* (Saint-Servais: Établissement d'éducation de l'Etat à Saint Servais, 1952), pp. 22–3; S. Pirard, *L'importance de l'observation dans la rééducation des jeunes délinquants. La méthode utilisée à l'établissement de Saint-Servais* (Namur: École sociale de Namur, 1949), p. 46.

36. D. Healy, *The Creation of Psychopharmacology* (Cambridge, MA: Harvard University Press, 2002); E. Shorter, *Before Prozac. The Troubled History of Mood Disorders in Psychiatry* (Oxford and New York, NY: Oxford University Press, 2009).

37. A comparison with the institutions for boys would show us whether the medical treatment of behavioural disorders was shaped by gender differences, like we suggest.

38. A. Wills, 'Delinquency, Masculinity and Citizenship in England 1950–1970', *Past & Present*, 187:1 (2005), pp. 157–85.

39. K. W. Tice, *Tales of Wayward Girls and Immoral Women: Case Records and the Professionalisation of Social Work* (Urbana, IL: University of Illinois Press, 1998).

40. M. Foucault, 'Le sujet et le pouvoir', *Dits et Écrits*, 4 vols (Paris: Gallimard, 1994), vol. 4, pp. 222–43; M. Ruchat, 'Observer et mesurer: quelle place pour l'infans dans le diagnostic médico-pédagogique? 1912–1958', *Revue d'histoire de l'enfance 'irrégulière'*, 11 (2009), pp. 53–73.

41. N. Rose, 'Government, Authority and Expertise in Advanced Liberalism', *Economy and Society*, 22:3 (1993), pp. 283–99; N. Rose, 'Governing 'Advanced' Liberal Democracies', in A. Barry, T. Osborne and N. Rose (eds), *Foucault And Political Reason: Liberalism, Neo-Liberalism and the Rationalities of Government* (London: Routledge, 1996), pp. 46–7.

11 Expertise and Trust in Dutch Individual Health Care

1. On the HPV vaccination campaign in the Netherlands, see P. Lips, 'Over de grens van wetenschap: de vaccinatie tegen baarmoederhalskanker', in H. Dijstelbloem and R. Hagendijk (eds), *Onzekerheid troef. Het betwiste gezag van de wetenschap* (Amsterdam: Van Gennep, 2011), pp. 75–95.
2. http://www.prikenbescherm.nl [accessed 25 January 2010].
3. *HPV Nieuws*, 19 March, 9 April and 17 December 2009.
4. http://www.nvkp.nl/ [accessed 25 January 2010].
5. http://www.niburu.nl/index.php?articleID=17538 [accessed 25 January 2010].
6. B. Goldacre, *Bad Pharma. How Drug Companies Mislead Doctors and Harm Patients* (London: Fourth Estate, 2012).
7. P. Hodgkin, 'Medicine, Postmodernism, and the End of Certainty', *British Medical Journal*, 313 (1996), pp. 1568–9.
8. S. P. Turner, *Liberal Democracy 3.0. Civil Society in an Age of Experts* (London: SAGE Publications, 2003); F. Huisman and H. Oosterhuis (eds), *Health and Citizenship: Political Cultures of Health in Modern Europe* (London: Pickering & Chatto, 2014).
9. A. de Swaan, *In Care of the State. State Formation and the Collectivization of Health Care, Education and Welfare in Europe and America During the Modern Era* (Amsterdam: University of Amsterdam, 1987); D. Porter (ed.), *The History of Public Health and the Modern State* (Amsterdam: Rodopi, 1994).
10. G. Weisz, *Chronic Disease in the Twentieth Century: A History* (Baltimore, MD: The Johns Hopkins University Press, 2014).
11. J. Le Fanu, *The Rise and Fall of Modern Medicine* (London: Abacus, 1999); C.E. Rosenberg, *Our Present Complaint. American Medicine, Then and Now* (Baltimore, MD: The Johns Hopkins University Press, 2007).
12. R. MacLeod (ed), *Government and Expertise. Specialists, Administrators and Professionals, 1860–1919* (Cambridge: Cambridge University Press, 1988); D. Pestre, 'Regimes of Knowledge Production in Society: Towards a More Political and Social Reading', *Minerva*, 41 (2003), pp. 245–61.
13. E. Ackerknecht, 'Anti-Contagionism between 1821 and 1867', *Bulletin of the History of Medicine*, 22 (1948), pp. 562–93.
14. See, for example, R. Evans, *Death in Hamburg. Society and Politics in the Cholera Years* (Oxford: Oxford University Press, 1987); S. Watts, *Epidemics and History. Disease, Power and Imperialism* (New Haven, CT: Yale University Press, 1997).
15. P. Baldwin, *Contagion and the State in Europe, 1830–1930* (Cambridge: Cambridge University Press, 1999).
16. Baldwin, *Contagion and the State*, p. 536.
17. D. Cannegieter, *150 jaar gezondheidswet* (Assen: Van Gorcum en Comp., 1954).
18. G. G. van der Hoeven (ed.), *De onuitgegeven parlementaire redevoeringen van mr. J.R. Thorbecke*, 6 vols (Groningen: Wolters, 1900–1910), vol. 6, p. 336.
19. See also: L. Ali Cohen, *Handboek der openbare gezondheidsregeling en der geneeskundige politie, met het oog op de behoeften en de wetgeving van Nederland*, 2 vols, (Groningen: Wolters, 1869–1872), vol. 2, p. 625.
20. M. Groen, *Het wetenschappelijk onderwijs in Nederland van 1815 tot 1980*, 9 vols (Eindhoven: University of Technology, 1983–1987), vol. 1, pp. 132–4; vol. 2, pp. 54–6 and vol. 7, pp. 65–8.
21. W. Stoeder, 'Brieven uit de hoofdstad', *Pharmaceutisch weekblad*, 11 December 1870;

15 January, 19 February, 19 March, 19 and 26 November, 10 and 24 December 1871
and 14 January 1872.

22. M. J. van Lieburg, 'De medische promoties aan de Nederlandse universiteiten (1815–
1899)', *Batavia Academica*, 5 (1987), pp. 1–17, esp. table 2.

23. J. W. Gunning, *Een blik op de toekomst der pharmacie in Nederland* (The Hague: De
Gebroeders Van Cleef, 1887), p. 3.

24. T. Rinsema, 'Brocades & Stheeman. Van apotheker-fabrikant tot farmaceutische indus-
trie', *Gewina*, 22 (1999), pp. 23–33.

25. *Verslag aan den koning van de bevindingen en handelingen van het Geneeskundig Staats-
toezigt in het jaar 1873* (The Hague: Van Weelden en Mingelen, 1874), p. 23.

26. [Petition concerning the Law on Medical Practice], National Archive The Hague (hereafter NA), Archief van de Tweede Kamer der Staten-Generaal, 1815–1945 (2.02.22),
inv.nr. 1252.

27. See R. O. van Holthe tot Echten (ed.), *De vrije uitoefening der geneeskunst of het artsenmonopolie?* (The Hague: Belinfante, 1913); J.A. van Hamel and E.C. van Leersum, *Vrije
uitoefening van de geneeskunde* (Baarn: Hollandia, 1914).

28. For a first batch of adhesions to the petition no. 39, see NA, archief Tweede Kamer
1815–1945, inv.nr. 1254, 1255 and 1518.

29. See, for example, *Weekblad van het recht*, 48 (1886), 5304.1; *Weekblad van het recht*, 73
(1911), 9240.3.

30. R. O. Van Holthe tot Echten, 'Why Should the State Protect *My* Health, *My* Life,
when My Health and My Life are in Danger only due *to My Own Will and My Own
Doing?*' in Van Holthe tot Echten (ed.), *De vrije uitoefening*, p. 49, italics in the original.
Van Hamel argued that current legislation was lacking in fairness, responsibility and
freedom: Van Hamel and Van Leersum, *Vrije uitoefening van de geneeskunde*, p. 20.

31. *Handelingen Tweede Kamer 1913/14*, pp. 1450–1451, 1454, 1464 and 1472–1473;
Handelingen Tweede Kamer 1913/14, bijlage A (appendix A), hfst. 5 (chapter 5), par.
12 (section 12), 3e afd. (paragraph 3), p. 6; *Handelingen Tweede Kamer 1913/14*, par.
13, 3e afd., p. 37.

32. J. den Hertog, *Cort van der Linden (1846–1935). Minister-president in oorlogstijd*
(Amsterdam: Boom, 2008).

33. NA, Ministerie van Binnenlandse Zaken, Volksgezondheid en Armwezen 1910–1918,
inv.nr. 417, 23 December 1916. The report is attached as Supplement I to the *Verslag
van de [juridische] Staatscommissie benoemd bij K.B. van 31 juli 1917 no. 39* (n.p.).

34. NA, Afdeling Volksgezondheid en Armwezen van het Ministerie van Binnenlandse
Zaken, 1910–1918 (2.04.54) inv.nr. 417.

35. *Verslag van de [juridische] Staatscommissie.*

36. NA, *Rapport van de [medische] Staatscommissie benoemd bij K.B. 31 juli 1917* (The
Hague, no date of publication).

37. NA, *Rapport van de [medische] Staatscommissie benoemd bij K.B. 31 juli 1917*, p. 33.

38. See, for example, 'Het rapport der commissie omtrent het onbevoegd uitoefenen der
geneeskunst', *Vox medicorum*, 2 April 1919, and 'Onbevoegd uitoefenen der geneeskunst', *Vox medicorum*, 28 May 1919.

39. *Nederlandsch tijdschrift voor geneeskunde*, 58 (1914), pp. 1846–2044 (30 May). All
further reference to the *NTvG* are to this special issue, unless otherwise stated.

40. G. Van Rijnberk, 'Een woord', *Nederlandsch tijdschrift voor geneeskunde,* 58 (1914), pp.
1846–2044, p. 1848.

41. R.O. van Holthe tot Echten, 'Het Nederlandsche tijdschrift voor geneeskunde over het art-

senmonopolie en de geneesvrijheid', *Het toekomstig leven*, 18 (1914), pp. 301–57, on p. 326.

42. See, for example, *Nieuwe Rotterdamsche Courant*, 27 September 1913; *De Telegraaf*, 11 October 1913; H. Treub, 'Vrije uitoefening der geneeskunst', *Vragen des tijds*, 39:1 (1913), pp. 155–72; A.W. van Renterghem, 'De vrije uitoefening der geneeskunst in Nederland', *De Gids*, 78:2 (1914), pp. 482–513 and *De Gids*, 78:3 (1914), pp. 74–103.

43. This trend was observed by contemporaries as well. See, for example, *Vox medicorum*, 31 October 1917.

44. *Verslag van de [juridische] Staatscommissie*, p. 5 (supplement I).

45. G.M. Bos, *Mr. S. van Houten. Analyse van zijn denkbeelden, voorafgegaan door een schets van zijn leven* (Purmerend: Muusses, 1952); S. Stuurman, *Wacht op onze daden. Het liberalisme en de vernieuwing van de Nederlandse staat* (Amsterdam: Bakker, 1992), chapter 5.

46. S. van Houten, *Staatkundige brieven* (n.p., 1913), 28 November 1913. For the same reasons, he was opposed to forced vaccination: NA, Papers of S. van Houten (2.21.026.06), inv.nr. 38.

47. M. J. L. A. Stassen, *Charles Ruys de Beerenbrouck. Edelman-staatsman 1873–1936* (Maastricht, 2000).

48. Stassen, *Charles Ruys de Beerenbrouck*, pp. 51–59. Ruys, who would be the leader of three Cabinets between 1918 and 1933, is characterized by his political biographer as 'the promoter of the social state': Stassen, *Charles Ruys de Beerenbrouck*, p. 197.

49. For the details, see *De Wet BIG en alternatieve geneeswijzen* (Amersfoort: Stichting IDAG / Corporation IDAH, 1992). Restricted interventions included surgery, anaesthesia and the use of radioactive material.

50. Le Fanu, *The Rise and Fall of Modern Medicine*.

51. F. Ankersmit and L. Klinkers (eds), *De tien plagen van de staat. De bedrijfsmatige overheid gewogen* (Amsterdam: Van Gennep, 2008).

52. T. Porter, *Trust in Numbers. The Pursuit of Objectivity in Science and Public Life* (Princeton, NJ: Princeton University Press, 1995); S. Timmermans and M. Berg, *The Gold Standard. The Challenge of Evidence-Based Medicine and Standardization in Health Care* (Philadelphia, PA: Temple University Press, 2003).

53. For the repercussions for the doctor-patient relationship, see E. Shorter, *Bedside Manners. The Troubled History of Doctors and Patients* (New York, NY: Simon and Schuster, 1985).

54. *De zorgverlening aan S.M. Een voorbeeldcasus.* (Report of the Health Care Inspectorate of The Netherlands) (The Hague: Staatsuitgeverij, 2004).

55. R. Braam, 'Voorwoord', in F. van Lunteren, B. Theunissen and R. Vermij (eds), *De opmars van deskundigen. Souffleurs van de samenleving* (Amsterdam: Amsterdam University Press, 2002), p. 7.

56. K. Wailoo et al. (eds), *Three Shots at Prevention. The HPV Vaccine and the Politics of Medicine's Simple Solutions* (Baltimore, MD: The Johns Hopkins University Press, 2010), p. 299.

57. Huisman and Oosterhuis (eds), *Health and Citizenship*. See also E. Tonkens, *Mondige burgers, getemde professionals. Marktwerking en professionaliteit in de publieke sector* (Amsterdam: Van Gennep, 2008).

58. U. Beck, *Risk Society: Towards a New Modernity* (London: Sage, 1992); A.R. Petersen and D. Lupton, *The New Public Health. Health and Self in the Age of Risk* (Sydney: Allen and Unwin; London: Sage, 1996).

59. Porter, *Trust in Numbers*.

60. W. Bijker, R. Bal and R. Hendriks, *The Paradox of Scientific Authority. The Role of Scien-*

tific Advice in Democracies (Cambridge, MA: MIT Press, 2009).

61. R. Pielke, *The Honest Broker. Making Sense of Science in Policy and Politics* (Cambridge: Cambridge University Press, 2007).

62. I would like to thank Willem Schinkel for his comments on an earlier draft of this chapter.

INDEX